Epitaph
for Planet
Earth

Epitaph for Planet Earth

How to Survive the Approaching End of the Human Species

DR. MILO DON APPLEMAN

FREDERICK FELL PUBLISHERS, INC.
New York, New York

For information address:
FREDERICK FELL PUBLISHERS, INC.
386 Park Avenue South
New York, New York 10016

Library of Congress Catalog Card Number: 81-70329

International Standard Book Number: 0-8119-0447-4

Manufactured in the United States of America

1 2 3 4 5 6 7 8 9 0

Published simultaneously in Canada by Fitzhenry & Whiteside United, Toronto

DEDICATION

*This book is dedicated to my fellow voyagers upon the planet Earth,
who hold in their hands
the potential for continued human existence
or, conversely, its destruction.*

Contents

Beauty is that Medusa's head
Which men go armed to seek and sever.
It is most deadly when most dead
And dead will stare and sting forever.

—Archibald MacLeish

The year is 2376 and the month is that called November when man last lived upon the earth. There is no longer the hum of a bee, the chirp of a bird, or the footstep of any carnivore, herbivore, or omnivore upon the surface of the earth. When God said I give you and your descendants this land forever and forever, who realized that he was speaking to the insects which burrowed in the earth and which themselves were destroyed when they left their safe refuge momentarily to examine the surface? Through pollution by wastes, by sound, and finally by radiation, man had managed to destroy all creatures that flew or that walked upon the surface of the earth. He was so proficient that even if larger creatures existed in outer space it would be a million years before they could inhabit this land.

The following is a history of how man destroyed himself and all other large creatures.

Preface

You may find the facts related in this book to be frightening. Good. You should be scared. We should all be scared to the point of action. Otherwise, we may read the epitaph and weep for generations of children who will never inherit the earth.

The problems discussed are not local, nor regional, in nature. Frequently, I use a specific city, or geographical area, for illustrative purposes. But fouled winds blow over national borders; polluted waters are shared by many countries. Exploding populations in one area cause repercussions thousands of miles away.

Let it be noted that I do not speak in this book of the potentials of a holocaust brought about by a nuclear war. Windborne radiation of atomic particles could move the epitaph up to the present time, or at least the next generation. Then all other problems discussed herein could become academic—unless a few isolated tribes should escape obliteration, or mutated monsters survive.

Life entails risk. Some risk we must accept with equanimity. We tend to become frightened of the wrong things—matters which actually are not threatening or which are insolvable. "A little learning is a dangerous thing"; and faddist prophets often beat their chests without a full understanding of the problems.

I have not written earlier for my fellow voyagers upon this planet. Previously, my writings have been directed to scientists; my lectures were delivered to physicians, pharmacists, and students of bacteriology and microbiology, here and abroad. Now, from the perspective of three score and ten years, with over five decades spent in this field, it seems time to speak directly to all concerned. If I do not, these warnings may go unheard, and the suggestions for remedies unheeded.

The Water We Drink

Water, water everywhere, nor any drop to drink.

—Samuel Taylor Coleridge

It is important that the sources of water utilizable by man be understood and, after that, why so much of the water is either foul or undrinkable. Almost forty years ago my wife had an artist paint a beautiful stream not far from Champaign, Illinois, where my pleasure had been to catch bass, bluegills, and blue channel catfish. The water in the picture is still clear and beautiful, reflecting the dancing sunbeams and capturing the reflection of the trees at the very spot I formerly fished. Three years ago I revisited this area. The trees for the most part had disappeared, the water was turbid and brown. An odor was present that informed me that the spillover or leachates from the cesspools or septic tanks of several camps and small industries had despoiled my favorite place. I was slightly nauseated; not at the odor, for in microbiology and bio-chemistry fouler odors were often encountered. It was due to the fact that this spot, once crystal clear, had been ruined as has been the water in so many other areas.

The people of earth are fortunate in that they live on the third planet from the sun, the water planet. On the second planet it would be found that the heat of the sun had converted the water into a heavy vapor cloud with probably no liquid on the surface. On the fourth planet from the sun the distance is so great from the source of heat that all of the water exists only as ice, rendering the surface unsuitable for man, animals, and plants.

Little of the water on our planet is available to man and most land animals, for even though 75 percent of the surface of the earth is covered by water, much of this cannot be used directly by man. In order to compare the water that is available with that available only with great difficulty, the easiest method of presentation is in tabular form. This will be one of the few tables in this book.

Total Water Available on Earth

Source	Approximate Cubic Miles	Approximate Percentage of Total Water
Oceans and seas	315,000,000	97.0
Glaciers and ice caps	6,000,000	2.15
Atmospheric moisture	30,000	0.01
Fresh and salt water lakes and inland seas	<30,000	<0.01
Average in rivers and streams	75 to 100	0.0001
Moisture of the Earth		
Soil moisture	<15,000	0.005
Ground water within 0.5 miles of surface	1,000,000	0.31
Deeper ground water	1,000,000	0.31

The table illustrates that less than 1.0 percent of the water of the earth is available to man from streams, rivers, lakes, and wells, if these sources are unpolluted—a factor that will be discussed in the next chapter. There is a surprisingly large amount of water that is completely unavailable. This is the water of crystallization of the minerals that comprise our earth. This is normally called the water of the lithosphere, which amounts to many times the amount of water in the oceans and the seas.

The water that exists in a frozen state, as glaciers and icecaps, covers approximately 11 percent of the land area, or 3 percent of the total surface of the earth. Were all of the icecaps and glaciers to melt simultaneously, the water level of the earth would rise about 150 feet. This would inundate almost all of the coastal areas of the world, and would fill the Mississippi Valley area and similar areas of most other countries.

Another illustration to compare the large amount of water with the small amount of land would be to imagine the earth as a perfect sphere. There would be no valleys, hills, or mountains, and there would be no depths of the oceans and lakes. If the water of these and of the ice were distributed uniformly on this sphere, it would be covered with water from 8000 to 9000 feet in depth. This is because the ocean attains depths of 35,000 feet or more, and has a mean depth of over 8000 feet.

It is astonishing how little fresh water is available to us and how carelessly we treat it. Where does the water come from that fills the lakes and keeps the streams and rivers flowing? It is easy to say "from rainfall," but where does this rainfall arise that keeps our land in some places so bountifully supplied and is in others so scant? If the sources of water were not renewable, man would not be alive, and were he ever to contaminate his area so badly that no water would be usable, he would be responsible for his own genocide.

The explanation for the circulation of the water that gushes down the mountain streams to rivers and lakes, and eventually to the ocean, is quite simple. The process is known as the *hydrological cycle*, or water cycle. It is a process that was suspected by Pollio Vetruvius, who was a Roman architect more than 1500 years ahead of his time. It was not until the middle of the seventeenth century, when Frenchmen such as Pierre Perrault and Edme Mariotte made simple quantitative observations, that the idea of the hydrological cycle was accepted.

Primarily, this cycle is based upon the fact that the sun evaporates water from the seas and oceans. In addition, it evaporates water from the land, and plants yield water to the atmosphere by transpiration. Although the following figures are not claimed to be accurate, the amount of water evaporated from the sea annually is approximately 95 million billion (quintillion) gallons, and that from plants and the land sixteen quintillion gallons throughout the entire world. Of this, about eighty-five quintillion gallons fall as rain at sea, and twenty-six quintillion gallons fall as rain on the earth.

Another way of stating this is that almost 400 billion acre-feet of water falls as rain or snow, of which three-fourths falls upon the ocean. The remainder of the water is formed into clouds and is carried over the lands to fall as rain or snow. One may be as philosophical as the Apocrypha, and ask:

Who can number the sands of the sea, and the drops of rain, and days of eternity.

The amount of water vapor in the atmosphere is greatest at the equator, least at the polar regions, and intermediate in the temperate zones. The amount of water that leaves the ocean by evaporation will balance out, and for this reason this is called a water cycle.

Many intermediate steps occur however. For example, primitive man, who was a hunter and fisher, used only two to three quarts of water per day. Modern man has developed a meat, cereal (bread), and vegetable culture. If we can assume that man eats a pound of beef per day, and the animal eats thirty pounds of alfalfa per day to yield eventually seven hundred pounds of beef, the water figures are startling. It requires approximately 2400 gallons of water per day to produce every twenty-five to thirty pounds of dry alfalfa. The beef steer drinks about eleven to twelve gallons of water per day, and it takes 300 gallons of water to produce a couple of pounds of bread. The human drinks only two or three quarts per day, but the total water requirement for the human for food and water is near 2700 gallons per day. In addition, every time man flushes a toilet, approximately three to six gallons of water are used.

What has happened to all of this water? All of it transpired into the air from the plants. It has entered the atmosphere, where it will again fall upon the earth or lakes or rivers. The water that is the wastes of animals or man either percolates into the topsoil, becomes runoff from the soil, or is carried by sewage systems for treatment and will eventually be returned to the ocean.

In the forty-eight contiguous states of the United States we are fortunate that we have rainfall averaging 5000 billion gallons per day. Of this amount, one-fourth is discharged by the rivers and the remainder by *evapotranspiration* from the land. The water that falls on saturated lands may cause erosion, carrying some of the soil with it to the streams and rivers as they rush towards the sea. These are ever-changing, a fact that was expressed over 750 years ago by Kamo NoChomei, the Japanese poet:

> The flow of the river is ceaseless and its water is never the same. The bubbles that float in the pools, now vanishing, now forming, are not of long duration.

One great difficulty is that water as rainfall is not uniformly distributed. Insofar as crops are concerned, it requires 1500, 4000 and 10,000 tons of water to produce a ton of wheat, rice, and cotton, respectively. Many areas of the world never receive this amount of water. The Bible and many other books speak of the scarcity of water in the desert, yet we know that Arab encampments and even great cities were founded in these areas by different methods. A message of hope, from Isaiah, as yet unfulfilled:

> The desert shall rejoice, and blossom as the rose.

One-half of the people of California live in the southern area of the state. The city of Los Angeles lies in a semiarid community. The rainfall on the average is scant, the city receiving 2 percent of the total rainfall of the state. The rainfall is also erratic, varying from furious storms to long periods of drought. Large numbers of persons moved into the area, but, unfortunately, they could not bring a water supply with them from their former homes. Water was always scarce in this area, and from the time of the great ranchos in the 1770's, men fought and died over who would own the waterholes and springs for a hundred years. As in many other semiarid states, battles were waged between the farmers and ranchers as to who would finally own the land.

By the early 1900's the city of Los Angeles realized that the local water supply was inadequate, and, by means that included chicanery, the city acquired control of the water of the Owens Valley watershed, which drained the east slope of the Sierra Nevadas to the north. The aqueduct was finished in 1913, but even as the water began to flow, the Metropolitan Water District planned to tap the Mono Lake watershed to the north, completing this in 1940.

Mono Lake has lost so much water that jagged rocks that were once completely covered now stand on barren land. The Navy had tried to blast channels so that coyotes could not cross and eat the eggs or young birds of the gulls, eared grebes, and the phalaropes that nest on the island. The migratory birds and their young are dependent on the brine shrimp of Mono Lake. Reports of July 1981 state that 97 percent of all baby gulls at Mono Lake have died. Starvation is the latest plague to attack the birds; since the city of Los Angeles diverts approximately 100,000 acre-feet of water from the water that would ordinarily flow into the lake, the hatch of brine shrimp upon which the birds would feed their young has fallen to only 10 percent of that of previous years. The Department of Water and Power of Los Angeles, of course, has disputed these figures. Los Angeles is now fighting for the Peripheral Canal that would remove water from the eastern fringe of the San Joaquin-Sacramento Delta. It would ensure the southern San Joaquin Valley and Southern California more water into the aqueduct, but it would imperil 700,000 acres of delta marsh lands vital to the migration of water fowl. It would also deplete the fresh water which flows through the delta into San Francisco Bay and flushes it of the destructive salt water that would intrude into this area from the Pacific Ocean.

Simultaneously, the city of Los Angeles built Parker Dam 155 miles south of Hoover Dam and impounded the water flowing down the western slopes of the Rocky Mountains at this point, bringing the water over 240 miles across the desert to the Metropolitan Water area. Recently, the Feather River Aqueduct was completed, which brings water from the north almost 570 miles to the Metropolitan Water area. On January 22, 1979, fears were rising concerning the amount of water being diverted to Los Angeles from Mono Lake, and almost simultaneously Governor Brown asked for an overhaul of the water laws of California.

Originally, Los Angeles had obtained its water primarily from *aquifers* (Latin for "water bearing"), which are underground layers of gravel, sand, or porous rock that can conduct the water any distance from mountain areas that have heavy rainfall or snowfall. If holes are dug or drilled to this strata of saturated sand or gravel, a well is produced.

In Texas, the Carrizo sand is a geological formation carrying water to an area twenty to fifty miles in width and hundreds of miles long. The sands range in thickness from 100 to 200 feet. Other formations are found in every state in this nation where water supplies are derived from wells, rather than from rivers or lakes.

The most unusual aquifers are the confined aquifers, a situation in which the porous rock, sand, or gravel has an impermeable layer of rock beneath it and an equally impermeable layer of rock above it. Thus the water-bearing strata are truly confined to a given layer. The upper end, or open end, of the aquifer lies in hills or mountains where it is continually charged with rain water. The water in the aquifer is under pressure, the amount of pressure being dependent upon the height from which it originates and the height of the ground at which the well is drilled. At one time, the city of Paris, France, secured a significant amount of water from an area known as the "Green Sands." At Artois the ground was quite low where the wells were drilled, so the water was under great pressure and gushed forth in fountains. These were given the name of Artesian wells from the locality.

The most unusual confined aquifers are those of northwestern Africa. These originate in the Atlas Mountains in an area of appreciable rainfall. Due to the height of the mountains the pressure on the water is considerable, and it cannot escape from between the top and bottom layers of impermeable rock except where cracks occur in the upper strata due to the forces of nature, or where man drills wells down into the permeable layers. Natural oases occur where the water oozes to the surface and whole cities depend upon this confined aquifer which reaches Libya. Algeria, a water-deficient desert country, could be transformed into a water-sufficient country by building one or more dams west of Algiers. It would be possible even to develop agricultural industry.

In southern California, one of the commercial companies that sells bottled water advertises that its product comes from a mountain spring of water that is pure and cold. This could be the perfection sought by the poet Horace in the following:

> . . . a piece of land not so very large, which could contain a garden,
> and near the house a spring of ever-flowing water, and beyond these
> a bit of wood.

Approximately three-fourths of the rainfall of the United States, some 2500 billion gallons per day, falls on half of the nation's area, where many of the crops are grown. Of this rainfall amount, perhaps 25 percent is lost as runoff. Water is gleaned more carefully from the watersheds of the mountains, however, and is delivered by aqueduct, not consumed in irrigation as practiced in the major areas of the Southwest.

Albuquerque, New Mexico, which rests on water-bearing sands, pumps 12,000 acre-feet of water through 118 thirsty wells annually.

Between 1964 and 1971 it completed the San Juan-Chamas project, which diverts water the city cannot use from the Rio Blanco, Navajo, and Little Navajo rivers into the Heron Reservoir in northern New Mexico. Meanwhile, the people of the Apache Jicarilla Reservation suffer because the loss of much of their water supply adversely affects the fish life and produces unpleasant tastes, odors, and turbidity in the water they are forced to use. This situation has been compared with the removal of water from Mono Lake by the city of Los Angeles.

Almost all major cities were initially built alongside a river or a lake. A good example from history is the city of Rome, founded beside the Tiber. As this river became grossly polluted, Rome had pure water brought by aqueducts from lakes in the hills and mountains, which were guarded by legionnaires from pollution. In the poorer sections of the city, wooden or leather plugs were abundantly spaced along the pipeline so that any person could withdraw a plug to fill a water vessel. The wealthy were permitted to divert the water through small pipes to fountains, usually in the atrium of the home, so that a constant supply of pure water was always available. All of the major satellite cities of Rome, whether in Italy, France, or Spain, had aqueducts conducting water to them. The author has seen these and marveled at the fact that when the pure water system of Rome was destroyed there would not be another for some 1300 or 1400 years.

Carthage, the bitter foe of Rome, did not have the large resources for a similar aqueduct system. It diverted to the capital all of the water available. In addition, every public building and all large homes were guttered so that during the rainy season the water would be collected and stored in vast underground caverns, which would hold sufficient water for twelve months.

At the present time, the cities of the United States are impounding large streams and rivers with dams in order to furnish their water supplies. Both New York City and Philadelphia receive their water supplies indirectly from the Delaware River. In 1965, there was a shortage of water in New York City, and simultaneously, incoming ocean tides almost reached Philadelphia with their brackish water. As a result it was necessary for New York City to release part of its impounded water in the hills in order to drive back this ocean water.

Man fears thirst. Unless extreme conservation measures are taken to conserve water, usage in the United States will double in the next fifty years. All fear a nation of dust and dryness. As T. S. Eliot mourned:

And the dead tree gives no shelter, the cricket no relief
And the dry stone no sound of water. Only
There is a shadow under this red rock,
(Come in under the shadow of this red rock)
And I will show you something different from either
Your shadow at morning striding behind you
Or your shadow at evening rising to meet you;
I will show you a handful of dust.

All cities have the same greed for water. Washington, D.C., fought to obtain 1200 million gallons of water per day from the Potomac River, which is more than the flow of the river. Naturally, this is an absurd and untenable situation, which matches closely the ideas and intellect of some of the men who make other regulations in this area. Sam Rayburn made a most acute remark:

The greatest domestic problem facing our country is saving our soil and our water. Our soil also belongs to unborn generations.

It is of great interest that two countries of the world in 1978 expressed their willingness to cooperate on a water-supply project. Iraq, through which the Tigris and Euphrates rivers run, has agreed to furnish Kuwait with fresh water through a massive pipeline which will be 320 kilometers in length. This pipeline should deliver 650 million gallons of water per day to Kuwait and will supply all needs projected to the year A.D. 2000. At the present time, Kuwait obtains its water from an expensive desalinization process. Only time will tell whether the overthrow of the Shah of Iran, and general unrest in the Middle East, will change this project.

Throughout the world, we are either overpumping our ground water supplies or impounding rivers, sometimes carelessly, in order to have water. During the lifetime of the author, the ground water level of Lucerne Valley of southern California has dropped from forty to 400 feet. In parts of the Mississippi River Basin the water has dropped 400 feet. Deep wells in parts of Arizona have dropped from 150 feet to more than 1500 feet.

Of the large amounts of water that are available daily, only 150 gallons per capita are used for household purposes. This includes the two or three quarts used as beverages. Most of the water is used in laundry, washing dishes, bathing, watering the lawns, washing the car, flushing the toilet, and miscellaneous areas. Approximately ten to fifteen times this amount is used per capita per day by agriculture, power plants, industry, and for other purposes. Almost all of this water that is used is returned to the hydrological cycle in one way or another, usually by sewage to be treated and disposed of in rivers eventually flowing to the sea.

In the United States, our daily usage of water, in billions of gallons, would be about as follows:

Use	Withdrawal	Consumed	Returned
Irrigation	160	130	30
Steam Power	120	2	118
Industrial	85	20	65
Municipal/Domestic	30	10	20

This table illustrates the fact that although the amount of water withdrawn from the hydrological cycle is large, almost all of this is returned at once. In industry where water is used for cooling, it is also almost all returned, whereas if it is used in processing of foods, it will be returned when the canned or frozen foods are eaten and the water is returned as treated sewage. In irrigation, apparently a larger amount is not returned, but actually this is not true. The growing crops use much water, but they lose this by transpiration of the leaves into the atmosphere where return is more indirect. Even the water that seeps into the ground is not lost or consumed, for the most part, as it goes into the build-up of the underground water table.

No matter what the source of water may be, it must be delivered to the consumer as a clear, sparkling, tasteless, and odorless product. It must be free of chemicals that are harmful, and of microorganisms and viruses that can cause disease.

On the other hand, man should fulfill his contract with nature not to waste water. For example, for each faucet that drips at one drop per second, not only are 115 gallons per month wasted, but the energy needed to pump the water from its source to the household is wasted. If this is from a hot water tap additional energy is lost. The true environmentalist in showering does not use twenty-five to fifty gallons of water, but either uses an automatic device to regulate flow or uses the taps on lowered pressure and showers quickly.

For those of us who live in older homes where each flush of the toilet uses three to five gallons of water, it is possible to place two or even three bricks in the tank to decrease the water used. Two-way flush controls are even simpler so that one can easily adjust the water used to the need presented. For those who conserve water a simple message is available:

> All your better deeds shall be in water writ, but this in marble.
> —Beaumont and Fletcher

The water we drink is water of the highest quality and is called potable water. There are other grades of water that are very useful to man, and some of these have more stringent requirements than the water that flows from our taps. The entire supplies of water might be listed as follows:

1. Potable water—drinking water for humans
2. Drinking water for cattle and other animals
3. Recreational waters
4. Water for washing and cleaning
5. Water for crop irrigation
6. Boiler waters
7. Cooling waters for canneries and power plants

The only waters that may have somewhat higher demands than potable waters are boiler waters and water for irrigation. Boiler waters must be free of permanent hardness that will build up "boiler scale," causing slow heat transfer or even breakage or leakage of the boilers. Frequently in areas of water with large amounts of permanent hardness, a hot water heater guaranteed for fifteen years will leak or rupture in ten years or less. That is because the insoluble salts of calcium, magnesium, and other metals present, even in good potable waters, precipitate to the bottom of the tank, forming a layer that heat penetrates through only with difficulty. If the homeowner were to open the bottom drain every six months and remove a bucketful of material containing the concentrated salts, his heater would outlast the guarantee.

In industry, boiler waters are treated with phosphates, silicates, and other compounds before the point at which the water comes in contact with heat. In this way no scale is formed in the boiler.

Irrigation waters in the southwestern states may contain alkaline salts which are often not observed by the homeowner who uses only small amounts. With the tremendous amounts of water used by crops, however, the salts or alkalis, particularly those of sodium and calcium, eventually can render the soil so alkaline that it grows crops only with difficulty. If the water contains much nitrite it can be toxic for man or animals.

Near industrial areas, sufficient sulfur dioxide may be discharged into the atmosphere where it can react with moisture, and eventually sulfuric acid is produced. It has been noted in Western Europe and Sweden that sufficient acid may be present in the rainfall to render the water unpalatable. In certain areas such as Lake Vanern in Sweden, the water may become very acid (pH 4.5) and the fish in the lake are destroyed. The same situations are being encountered in the smelter areas of the United States.

All waters that are to become the waters we drink should be freed of any suspended matter that interferes with the clarity of the water and should be treated to destroy bacteria and other microorganisms that are harmful to man. It is possible to use these waters as recreational waters if one so wishes and remembers that the waters will be polluted. For example, the water of the Feather River Aqueduct in California is used for a series of recreational lakes for boating and fishing. After the water leaves these lakes, however, it should be impounded in reservoirs for thirty to sixty days to destroy the cysts of the organism causing amebic dysentery, which are difficult to kill with chemicals. If algae start to grow in the reservoirs they can be destroyed with treatments of very dilute copper sulfate.

The water from the reservoirs is eventually conducted to large mixing tanks where a very dilute solution of alum is added. Into this a small amount of lime is stirred, which forms a very flocculent and sticky solution of aluminum hydroxide to which insoluble impurities will adhere. This mixture is then filtered and a bright, sparkling water is obtained. To destroy any possible disease-causing bacteria that have not been removed, 0.2 parts per million of chlorine are added. This amount is tasteless and only when it is exceeded, or external water pipes are being treated, is there a chlorine taste or odor in the water. It should be noted that currently claims are being made that the combination of chlorine with certain chemicals produces water that is "carcinogenic." Data are now being gathered from New Orleans and other cities along the lower Mississippi to test this premise. The Council on Environmental Quality has linked the usage of chlorinated water in some areas with increases of from 10 to 90 percent in rectal cancer and an increase of 50 percent in cancer of the bladder. All waters should be treated to coagulate, filter, or destroy organic compounds before the use of chlorine, chlorine dioxide, or ozone to prevent the formation of carcinogenic hydrocarbons.

No water is absolutely pure in being free of chemicals, including salts, unless it is freshly distilled water in the laboratory. So-called natural waters are not necessarily pure. Even rainwater falling through the air first picks up dust and later dissolves carbon dioxide of the air. Except for freshly distilled water, all others contain salts, many to the extent that they cannot be utilized. The following illustrates salt concentrations in water from various sources:

Water Sources	Salts (Parts Per Million)
Freshly distilled water	0
Lake Tahoe	70
Lake Michigan	170
Missouri River	360
Pecos River	2600
Ocean	35,000
Dead Sea	250,000
Great Salt Lake	260,000

Salts or salinity may not only be present in water sources and render them unusable, but the salts may tend to ruin soils. In Pakistan unlined irrigation ditches had been used for three-quarters of a century. The land is flat and primitive agricultural methods were used. As a result, the salts present in the soil were at first concentrated due to the leakage or leaching of the soil. Then, since sufficient water could not be obtained, the salts traveled upwards in the soil as capillary moisture. The concentration of salts was such that crops were diminished or destroyed. In 1958, Pakistan tried to correct the problem by digging deep drainage ditches and applying large amounts of water to the surface of the soil to drive the salts downwards where the salty water would be carried away. This method is succeeding to some extent.

In the United States, the Clean Water Act of 1977 amended the previous act of 1972 and, as a result, many sources of pollution have been controlled or are being investigated. This act includes many very important points and will be discussed in the next chapter on water pollution

How We Pollute Our Waters

The purity and safety of our drinking waters are threatened by two types of pollution. The first of these is called *non-point* pollution because it does not arise from specified sources such as sewage disposal plants or a specified industry, but represents the damage caused by fertilizer and soil runoff in farming, pesticides from the spraying of crops, and other miscellaneous sources. This type of pollution is approximately 300 times greater than that discharged from municipal sewage plants and industry, which are called *point* or specific sources. Both sources will be discussed in this chapter.

Together they render 10 to 15 percent of our water hazardous to health, as this amount may contain toxic chemicals, pesticides, alkalis, radioactive or solvent materials, or disease-producing organisms. In many or most areas of the country, tests are used to determine if poisonous or carcinogenic materials such as organic or inorganic solvents, polychlorinated biphenyls, or other similar materials are present. The health departments of many areas test solely for disease-producing organisms or bacteria that indicate their possible presence, and for heavy metals such as mercury, lead, and cadmium. Tests for these are not done in other locations, particularly in smaller communities, and to drink many waters brings to mind Wilson Mizner's comment on Hollywood: "It's a trip through a sewer in a glass-bottomed boat."

Since non-point sources of pollution are more frequent than those from specific sources, a close look should be taken as to how this type of pollution occurs. Pesticides will be discussed primarily in a separate chapter since their effects can be so far-reaching where fish, birds, and animals other than man are concerned.

In 1971, I read an article to my class in Man and His Environment that related that 200 shells filled with the most deadly nerve gas had been stored on the ice of a frozen Alaskan lake in 1966. The shells had been forgotten and had sunk to the bottom of the lake in the spring thaw. It was not until summer of 1969 that the shells containing VX nerve gas, of which one drop on the skin is fatal, were supposedly recovered and the poison neutralized. Not one student in the class seemed to be impressed with that fact, nor particularly frightened by the awareness that our water supplies are being continually contaminated from one source or another. In May 1969, other gas bombs, in leaky condition, were found in Colorado. The governor of Colorado wishes these to be moved to Utah; the governor of Utah is not in agreement—but the Army plans to move these to the Tooele Army Depot in the western Utah desert.

A lake or lakes can be used to illustrate the non-point sources of pollution, although the same factors apply to streams and rivers. Pure, young mountain lakes are known to be *oligotrophic*. They contain little organic matter washed in from their shores and as a result the fish supply in the clear, cold lakes is limited. The fish will tend to grow quite large as the species prey upon one another. The small amounts of nitrogen and phosphorus have limited the amount of small plants called *phytoplankton*, upon which the smallest animal forms called *zooplankton* feed. As a result, the population is self-limiting.

In a similar lake surrounded by a farming community the land has been heavily fertilized with nitrogen and phosphate fertilizers to produce large crops. As a result, during heavy rainfall topsoil, including part of this nitrogen and phosphate, will be washed into the lake. The algae or phytoplankton will grow rapidly, so much so that in summertime they may produce a scum on the surface of the water. The small animal forms that eat this algae may grow abundantly unless there are limiting circumstances that do not permit them to grow.

The limiting circumstance of greatest importance is oxygen, for all fish, and most other animal forms, require oxygen. To understand how oxygen can disappear from the waters in late spring and late fall one must realize that water is at its heaviest at 39.2 degrees Fahrenheit or 4 degrees Centigrade. During the summertime the water at the surface of a lake rises to or near air temperature and forms a layer of water that is many feet thick and much lighter than the colder water below. Oxygen of the air dissolves in this water and the insects and other forms of life that fish use as food thrive. There is abundant food for fish, so usually more fish are found near the surface.

Nearer the bottom of the lake there are also abundant nutrients, but since there is little circulation between the top and bottom waters, microorganisms such as bacteria use the nutrients and also scavenge this area for oxygen. This water becomes stagnant and fish enter this area infrequently for short times to feed on bottom growth.

In the fall, as winter approaches, in temperate climates the surface waters cool to 4 degrees Centigrade (their heaviest) and sink to the bottom. This results in a great overturn of the waters. The nutrients present in or on the mud circulate to the top, and the whole lake now has oxygen as these waters are exposed to the air. If the lake becomes so cold that it freezes over, there is little activity of microorganisms at either the top or the bottom. Consequently, little of the oxygen at the bottom, or top, of the lake is used up.

When spring approaches and the waters at the top reach 4 degrees Centigrade again, there is a recirculation of the waters of the bottom and the top known as the "spring overturn," in contrast to the previous "winter overturn." The waters are saturated with nutrients, most of which have been carried in with the spring rains; and, as the waters are warmer at the surface and sunlight penetrates a number of feet, the green algae plants called phytoplankton grow in profusion. These green plants produce oxygen using carbon dioxide of the air and furnish food materials for the small animal forms (zooplankton) upon which the fish may feed. If too much fertilizer material has been carried into the lake, the elements such as nitrogen and phosphorus may be so abundant that the waters become stagnant. Bad odors develop and the entire process is called *eutrophication*, or "overfertilization." As Coleridge said of one area:

> I counted two-and-seventy stenches,
> All well defined and several stinks.

In discussing water sources it is necessary to refer briefly to contamination of water. Turbidity is one type of non-point source of contamination that may occur as runoff from the soil after a rain carries particles of silt and clay into a river or a lake. If the land has been manured, in all likelihood more particulates will enter the water. If the water is very turbid, sunlight will not penetrate and, as a result, photosynthesis and growth of algae will be minimal. The particles tend to settle out and in time become part of the river or lake bed. Silt which has larger particles than clay will settle out first. Turbidity may also result from the effluents of the steel industries, mines, and sewage, which will be discussed as a separate "point" topic later.

These effluents and soil usually carry appreciable amounts of nitrogen and phosphorus compounds. It is these compounds that are the support of the process of eutrophication which can damage or destroy a lake through several processes. It only requires from 0.02 to 0.05 parts per million (ppm) of phosphorus and from 0.2 to 1.0 ppm of nitrogen to start a heavy growth of algae, which is known as an algal bloom. It makes no difference whether the phosphorus and nitrogen came from the soil, from industrial wastes, or from other sources such as sewage or detergents.

Normally the detergents from point sources consist of three parts: a *surfactant* or sudsing agent such as a phosphate; a corrosion inhibitor which may be a silicate; and a compound, such as carboxy-methyl-cellulose, that prevents the redeposit of dirt, holding it in the water.

In any case, when sufficient phosphorus and nitrogen enter the water the algae multiply rapidly, using sunlight as their source of energy and carbon dioxide dissolved in the water as their source of carbon. This growth may be so luxuriant that it will form a green or greenish blue scum on the water, depending on the type of algae present. Many of these algae give obnoxious tastes and odors to the water. In addition, certain protozoa which feed upon these may also give foul odors. In this way a lake that was at one time clear and usable may become foul. Thoreau well stated:

There is no odor so bad as that which arises from goodness tainted.

These are only a few of the processes that occur in eutrophication and that have resulted in the death of many lakes, and, with other causes, almost destroyed Lake Erie. As the algae and protozoa die they are digested by many microorganisms, including bacteria. Some of the microorganisms, such as *Streptomyces*, can give the water an earthy taste and an odor similar to that of a newly plowed field. The microorganisms use up the oxygen of the water in digesting the algae, protozoa, or even weeds that have grown. Without oxygen fish cannot live and massive fish kills have resulted from this deoxygenation.

An almost total destruction of the oyster industry in the Great South Bay, which lies between the south shore of Long Island and Fire Island, occurred due to eutrophication. The streams that flowed into the bay were lined with farms on which three million ducks per year were raised. The fecal material of the ducks entered the streams and was carried into the bay. This is primarily "point" source pollution. The water is brackish, for not only does ocean water enter through two narrow openings, but the polluted "fresh" water flowing in mixes with this, lowering the salt content. Oysters feed upon an algae called *Nitzchia*, which grew in abundance before all of the duck farms were established. As the water became turbid and eutrophic in some places, the *Nitzchia* were destroyed and replaced by a very tiny algae, *Nannochloris*, which the oysters could not use, and this industry was destroyed. Interestingly, clams can use this latter algae and have replaced the oysters. The duck farms were ordered to clean up their effluent before discharging it into the streams.

Florida had a great problem with the growth of aquatic plants in the inland waterways. The most difficult of these to deal with were the water hyacinths that had been accidentally introduced. In 1965, the state imported manatees (sea cows) to assist in eliminating the weeds, but, although they do a good job in the destruction of the weeds through grazing, they reproduce too slowly to solve the problem. In 1977, two biologists of the California Department of Food and Agriculture, who were fishing in the All-American Canal for catfish near Calexico, noted that the canal was becoming fouled with hydrilla, a rapidly spreading plant from South America. As in Florida, this plant grows rapidly, then dies, and bacterial decomposition deoxygenates the water, destroying fish. As of 1978, approximately 380 miles of the canal had been closed to fishing and it is hoped that this noxious invader can be destroyed.

Iron bacteria are organisms that may contaminate reservoirs and lakes and, although they are not harmful to health, they may be nuisance organisms to the water companies and to the homemaker.

With the exception of one group of iron bacteria called *Gallionella*, all of the other groups need organic matter. All of the iron bacteria have the ability to convert ferrous iron, which is soluble and when very dilute is colorless, into ferric iron, which is reddish brown. There are many instances where ferrous iron is present in the water supply and everyone is completely unaware of its presence until the iron bacteria, which were originally few in number, are stimulated to multiply and oxidize it to the ferric form. Almost overnight, it seems, the water turns reddish brown; and, if this has not been carefully monitored, the water department receives hundreds of calls from irate homemakers that their white clothes have large brownish spots. Since these are actually rust spots, they are quite difficult to remove. This non-point source of pollution is troublesome.

In coal mining areas, a combination of difficulties is frequently encountered. In parts of West Virginia, Pennsylvania, Ohio, Kentucky, Tennessee, and other states where coal is mined, the coal contains yellow spots that are called "fool's gold." This is iron sulfide, or iron pyrite, which is iron sulfide. In the water of the mines, as in abandoned mines, bacteria that are capable of oxidizing sulfide to sulfuric acid produce larger concentrations of the acid. The iron also may be solubilized. With the spring rains the sulfuric acid is washed into streams and rivers, where it can kill fish. As the acid is diluted out, the iron-oxidizing bacteria can produce the red-brown discolorations. Actually, the above is a mixture of point and non-point contamination.

This same problem has been encountered in Britain, France, Poland, and Germany. All abandoned mines should be sealed thoroughly so that these acid waters cannot leach out. This is difficult to do since even concrete seals may eventually be corroded.

If the water containing the sulfuric acid or sulfates reaches a stagnant area where organic matter is present, other bacteria, *Desulfovibrio*, can reduce the sulfate to foul-smelling sulfides. These gases are particularly obnoxious, so much so that traces of them are added to natural or artificial gases furnished to the households or industries. Thus, a gas leak can be detected by odor immediately and persons are not overcome by natural gas, which has little or no odor.

The sulfides, whether formed from sulfates leaching from mines or from the wastes of tanneries, steel mills, or other industries entering streams and rivers, can result in massive fish kills because the source of oxygen has disappeared. The windrows of dead fish one sees beside and upon the lake surface were those that were dependent upon oxygen ordinarily present in their accustomed waters. Nine milligrams of oxygen per liter is the saturation point. Five to seven milligrams of oxygen, however, is average, and satisfactory for fish life under most conditions. When the oxygen falls below four milligrams per liter, this level indicates grave pollution. If these fish could speak for themselves, the great Shakespeare might describe one of these:

> Who lin'd himself with hope, eating the air on promise of supply.

Upon occasion, salinity may be so great in waters that they cannot be useful in a conventional sense. This is not limited to the Great Salt Lake, the Dead Sea, or similar bodies. One great source that could be stopped is the brine that is pumped from oil wells and permitted to flow into streams or rivers. In the Los Angeles area these saline masses usually enter sloughs that flow into the ocean. Along with the salt brine, hydrocarbons and other toxic compounds are also carried.

In 1982 we are now becoming aware of the fact that Mono Lake in California has been ravished of its water supply due to the thirst of Los Angeles for more and more water. Fresh water flowing into the lake was diverted from it. The water has become several times saltier than that of the ocean and the formerly magnificent major bird habitat may be destroyed, as the northern shore has become a peninsula which can be invaded by land animals. Thus, the breeding ground of the seagulls that feed upon the brine shrimp will be destroyed.

There are other toxic compounds present in water that are more insidious than those previously mentioned. Elsewhere* we discuss pesticides and herbicides, which are usually non-point pollutants. Not only can they cause death in fish and birds, but these as well as other pollutants are carcinogens; that is, they have the ability to produce different types of tumors or neoplasms in fish. When one realizes that certain cities such as New Orleans must use water that has been contaminated multiple times by the cities 500 to 1000 miles above it, one can only be thankful or hopeful that his water supplies supposedly come from better sources.

The following carcinogens (cancer-producing agents) have been found in the drinking water of New Orleans and Carville, Louisiana:

---

*See Chapter 6, "The Cides That Kill or Injure Us," page 98.

aldrin, dieldrin, DDT, benzene, and carbon tetrachloride. A much greater amount was present in the untreated water before treatment. It has been theorized, however, that chlorine, used for purification, may combine with certain pollutants to aggravate the carcinogenic properties. Many cities in Missouri, Iowa, Nebraska, and Ohio are exposed to some or most of these water-carried cancer- or leukemia-producing agents. Some of these cities have a 25 to 35 percent greater cancer mortality rate than the national average.

There are at least a thousand organic compounds or heavy metal compounds that can enter the water through sewage or industrial wastes. We are all aware of the current dramatic pollution problem caused by the drums of toxic chemicals buried decades ago, which are now disintegrating and destroying communities. There are at least 40,000 to 50,000 areas in the United States where toxic chemicals have been dumped. At one time this was not illegal, but since laws now control the dumping of toxic wastes, these can be disposed of illegally in dumps or along roadsides and in the woods. Many of these compounds can leach into the underground water supply where they contaminate the water system of a farm, village, or even an entire city.

Fortunately, courts are taking gross pollution seriously. In June 1981 a businessman of Raleigh, North Carolina, was fined $209,000 and sentenced to thirty days in jail, and his employee was sentenced to eighteen months in prison for dumping toxic PCB (polychlorinated biphenyl) alongside roads in ten counties.

In August 1981, a contract for $1.97 million was signed by the Paul Hubbs Construction Company and the Santa Ana Regional Water Control Board to clean up the Stringfellow hazardous waste site, which has threatened downstream and groundwater pollution. Almost at the same time, employees of an industrial plant were returning to work following the spill of over 1000 gallons of silicon tetrachloride, which produces the highly toxic hydrochloric acid. Twenty-eight persons were injured in the latter accident, in addition to property damage.

Unfortunately, the Environmental Protection Agency (EPA) in the latter part of 1981 is considering whether or not toxic dump operators should be permitted to drop their insurance against toxic spills. If this does occur, it would be catastrophic as no central control would be exerted.

One industrial carcinogen must be mentioned at this point since it is appearing in wells all over the United States. *Trichloroethylene*, a solvent and grease-cutter used in industry and in septic tanks, was noted in well waters in Bucks County, Pennsylvania, in September 1979. Almost simultaneously it was found near Sacramento; since it is a carcinogen, a warning was given. The source of the last discharge was from Aerojet General Corporation's 8000-acre industrial complex. Aerojet had been disposing of the TCE in a pond that, by law, was only permitted to contain the wastes of septic tanks. In January 1980, a family in Rahns, Pennsylvania, experienced difficulties. The skin of the wife itched, and the laundry turned red when chlorine was added. The TCE content of their water was 300 times the level considered safe. The source undoubtedly was a rod and wire manufacturing firm known as Techalloy.

In the Los Angeles County area, TCE pollution did not close wells in East San Gabriel Valley January 9, 1980, although the concentrations reached five parts per billion; but on January 16, 1980, one of the wells was closed because of its high level. Other wells were closed on January 17 in San Gabriel Valley; on January 19 at Atwater and North Hollywood; on January 22 at Temple City; on January 26 at Santa Monica; on January 29 at a dairy, an ice plant, and a home in the San Gabriel Valley; and ad infinitum, all in 1980.

In Florida in January 1980, a study of the state's drinking water showed that fifty-nine major water systems contained chemical carcinogens, and in Philadelphia, the EPA found eight cancer-causing chemicals in levels above the proposed federal standards.

In Montague, Michigan, in September 1979, the water wells of the area were contaminated with the toxic chemical C-56, or *hexachlorocyclopentadiene*, a chemical used in many pesticides. The company responsible for the contamination was hostile, but eventually connected every home that had its own well to the city water supply. This same curative plan was followed, some years earlier, when companies storing gas in supposedly impervious strata near Herscher, Illinois, destroyed the farm wells in this area. In Sacramento, in mid-December, the federal government and the state of California filed a multimillion dollar suit against the Hooker Chemical Corporation and Occidental Petroleum Corporation for contaminating the wells near Lathrop, California, with suspected cancer-causing chemicals.

In Rome, Italy, the Tiber River is grossly contaminated. The newspaper *Paese Sera* stated, "The Tiber kills because it has been abandoned; amid virus, rats and garbage it awaits a renaissance."

Toxic chemicals may be added to our water supplies as the result of non-point pollution. Rainfall is leaching radioactive materials left in the waste heaps from the mining of uranium, and is entering the water supplies. Rain also washes nitrite and nitrate from farmlands into the waters of streams or rivers, or deep into the ground where these compounds contaminate wells. Residuals of cadmium, chromium, lead, and mercury from the natural ores in which these compounds exist enter the waters. Cyanide, fluoride, selenium, and, to a lesser extent, arsenic and manganese, are usually derived from point sources such as manufacturing plants or sewage disposal plants.

Of the toxic chemicals mentioned, the outbreaks of mercury poisoning are most spectacular and will be discussed in greatest detail. The sources of mercury poisoning include erosion of natural deposits of cinnabar (the common mercury ore), a non-point type of pollution which may be exposed or submerged, and the washing of mercury wastes from mine deposits into the water supplies. Much greater amounts of mercury have been used in the past century and mercury poisoning has increased as a result.

Mercury may be carried into the atmosphere and precipitated out by the moisture of the air. In this way it has been possible to measure the amount of mercury over the last 2800 years by measuring the amount of mercury at different depths of the Greenland Icecap and the Antarctic. This indicates the levels at a given period of time. Mercury was present in approximately the same amounts in layers of ice that represented 800 B.C., A.D. 1714, A.D. 1815, and A.D. 1946. All were approximately thirty to seventy-five parts per billion. The early deposits represented degassing of the earth's crust and these amounts did not vary greatly. Beginning approximately in 1950, the amounts doubled and tripled due to mercury liberated from fossil fuels, and from cement and other industries, where the temperatures reach such a high level that this metal is liberated from the fuel, limestone, and shale in which it occurs in small amounts.

When the point sources of possible mercury poisoning are considered, of the world production of mercury (which is 10,000 tons per year) approximately one-third of this amount is used in the United States, as shown on page 24.

Mercury serves as a catalyst in the electrolytic production of chlorine and an appreciable amount is lost to the environment, where it can enter rivers, the atmosphere being rendered toxic in any event.

Industry	Approximate Tons Mercury Used Annually
Electrolytic production of chlorine	1500
Electrical Industries	1400
Paint Industry	740
Instruments	400
Catalysts	230
Dental	200
Agricultural uses	200
General Laboratory uses	125
Pharmaceutical	50
Pulp and Paper Industries	40

This liquid metal is also used in silent switches in electrical industries, in addition to other uses. When these are discarded, the mercury can enter the environment. Since mercury compounds are highly toxic, they are used in the paint industries to prevent mildew and other attacks by microorganisms upon wood and various substrates, including the painting of ship bottoms. Mercury is an important component of many scientific instruments, including thermometers and electrodes. As a catalyst, mercury is used in the production of vinyl chloride. The great outbreaks of mercury poisoning that alerted the people to its toxicity in fish and shellfish arose in this manner, and will be discussed under the toxicities of Minamata Bay in Japan.

Mercury forms an amalgam when mixed with silver. There are few people in western countries who do not have silver amalgam fillings in at least a few of their teeth. The toxicity of mercury as an amalgam is probably negligible, although it is possible that microorganisms that form plaques upon the teeth could convert at least some mercury to toxic form.

A great and dangerous use of mercury has been in the agricultural industries. Seeds are treated with mercury preparations to prevent the attack of microorganisms upon the sprouting seeds and bulbs. At one time, bichloride of mercury ($HgCl_2$) was used, which was not as dangerous as organic mercuricals that developed later.

In the laboratory mercury has had many uses. One of these was to act as a catalyst in the determination of protein in food products by a method called the Kjeldahl process. In more recent years, mercury has been replaced in part by copper, which is relatively nontoxic.

Pharmaceutical industries use fewer and fewer mercury compounds. Calomel was used frequently at one time, but has been replaced by better and less toxic compounds. Mercury is still used, however, in the pulp and paper industries as a slimicide. That is, the mercury compounds that are used destroy bacteria and other microorganisms which attack the wood pulp and paper substrates. Since this mercury compound enters the streams and rivers directly, such use should be prohibited.

It has been known for centuries that mercury is toxic, and many suspect that Ivan the Terrible of Russia, Charles II of England, and Napoleon, as well as many lesser-known victims, were eliminated with mercury salts. In 1953, a number of cases of mercury poisoning occurred in Japan, with over a hundred deaths or irreparable injuries from eating raw or uncooked fish or shellfish obtained from Minamata Bay.

Both acute and subacute poisonings were observed. The victims usually showed symptoms first in the extremities, lips, and tongue. They developed an ataxic gait, clumsiness of the hands, blurring of the vision, and deafness. In children the brain was affected. There was an ashen-gray appearance of the mouth and pharynx, gastric pain, and severe and bloody diarrhea. Fish-eating birds, cats, and other domestic animals were also poisoned.

This poisoning resulted from a factory bordering Minamata Bay that produced vinyl chloride. In this process acetylene gas in the presence of hydrochloric acid is converted to vinyl chloride, using mercuric chloride as a catalyst. For each ton of vinyl chloride produced, somewhat over two ounces of mercuric chloride is lost in the waste water.

This sounds like a minimal amount, but in each year two tons of mercury were lost. Although this settled into the mud of Minamata Bay, microorganisms were present that were able to convert it to the highly toxic organic forms such as methyl mercury, dimethyl mercury, and phenyl mercury compounds. Other highly toxic organic mercury compounds were also produced, all of which were able to bind to the proteins of cell membranes and produce many terrible changes.

In 1965 at Nagata, Japan, another outbreak occurred from similar causes which poisoned 120 persons. This involved only the fish and shellfish of the Agano River, where discharges of mercury were made. There were no cases amongst the fisheries in the nearby coastal areas. Rupert Brooke stated one attitude clearly:

And in that Heaven of all their wish,
There shall be no more land, say fish.

The first outbreaks had alerted the Swedes to study the problem of mercury and to call an international symposium in 1966. As early as 1950 Swedish bird watchers had noted high death rates of wild birds such as pigeons, pheasants, and partridges caused by the methyl-mercury coating of seeds. They eventually knew of the epidemic poisonings among the people in Iraq in 1956, and in West Pakistan in 1961, which were due to the sale and use of mercury-treated seeds. In 1965 it was proven in Sweden that the increasing amounts of methyl mercury in the eggs of wild birds and hens were associated with similar rising levels in the meat of both swine and beef cattle and these, in turn, correlated with the mercury used in seed treatment and in paper mills.

In the United States, farmers were using between thirty to forty times as much mercury per acre as in Sweden, but for a time these levels failed to cause alarm. The nonindustrial or farm usages are probably non-point sources of poisoning.

The Swedes made a consistent search for the dangers of mercury and the rising levels. The earliest seed treatments had been made with inorganic mercury salts that were not as toxic, but in the 1930's a change was made to phenyl mercury compounds, and in the 1940's to alkyl mercury compounds. These latter not only are more toxic in their own right, but also are readily converted to methyl mercury, which is highly toxic. They also studied feathers removed from birds in museums that had been prepared as early as 1840 and compared these with the feathers of similar species at the time of their investigation. A well-documented increase in the mercury in feathers of the modern birds was noted, which was up to twenty times as high as that for birds from the earlier period. The Swedes worked out a formula so that they could estimate the mercury in the flesh of birds at these different times. They also realized that fish from certain areas were unsafe because of the amount of mercury they contained. The scientists of that country were the first to prove that seeds of plants grown from methyl-mercury-treated seeds contained over twice as much mercury as seeds from untreated plants.

In California, treated seed is frequently eaten by game birds, and in 1969 it was shown that pheasants had a level of 1.4 to 4.7 parts per million of mercury in the tissues. In certain states it was legal to hunt but not eat the killed birds because of possible toxicity. In Alberta, Canada, hunting was canceled when a level of 1.0 ppm was found in the flesh of pheasants. It should be noted that in 1969 three children of one family in Alamogordo, New Mexico, were poisoned by eating the flesh of hogs that had been illegally fed mercury-treated seed. Physicians at first did not recognize mercury poisoning in these cases, probably because our practice had not kept up with the known literature of other countries.

As earlier stated, organic mercury compounds, particularly methyl mercury, have been used as fungicides and slimicides in the wood pulp and paper industries for years. Although the permissible amount was decreased in the United States, a sufficient amount could enter waters so that algae, the zooplankton feeding on algae, and fish could accumulate amounts past those regarded as permissible. Although the World Health Organization would prefer to see the permissible amounts in fish decreased to 0.05 parts per million, the Food and Drug Administration permits a level up to 0.5 ppm, ten times that amount.

Since mercury is a heavy metal, it accumulates in the body. Thus, larger fish that prey upon smaller fish that might have permissible amounts of mercury can accumulate amounts that are far above the permissible levels. For some time difficulty was encountered with both tuna and swordfish. Fish-eating birds also accumulate mercury in the body, which can be passed into their eggs, rendering them nonviable. As Pope wrote:

> Oft, as in airy rings they skim the heath,
> The clam'rous lapwings feel the leaden death;
> Oft, as the mounting larks their notes prepare,
> They fall, and leave their little lives in air.

Although the specific sources of mercury from industrial outlets have been mentioned, there are two sources that can add significant amounts of mercury to waters which can eventually reach the sea. The first of these is the natural erosion of cinnabar, which is the ore from which mercury is mined and processed. This erosion has continued for countless ages and accounts for much of the mercury present in the oceans and seas.

Another source is the burning of coal, which releases mercury into the atmosphere. The amount of mercury released in this way has steadily increased with the usage of this fuel and at present is approximately 2500 to 3000 tons of mercury worldwide.

Death from lead may be due to many sources, most of which are not water borne. Hippocrates in the fourth century B.C. described the toxic effects of lead colic in a miner. Slave labor was usually used by the Romans to mine lead since the life span of the worker was limited. A certain amount of lead poisoning occurred in the major Roman cities since lead pipes were used to conduct the water from the aqueducts to the homes of the wealthy. In addition, copper drinking vessels were lined with lead which dissolved either in water or in wine. Lead glazes on pottery were, and are, dangerous to health.

Lead carbonate or bicarbonate was solubilized in the water in small amounts and undoubtedly some cases of lead poisoning resulted. This was not as significant as in England in the 1700's and early 1800's, where water was pumped from cisterns through lead pipes to tanks that were, in turn, lead-lined to proof them against leakage. In certain households many family members died of lead poisoning since lead, like mercury and other heavy metals, has a cumulative effect. Lead may have many effects, among which is permanent brain damage. Lead poisoning can also cause sterility in men.

It is interesting that the amount of lead in the atmosphere has been measured in the northern hemisphere and we know the amount deposited over a long period of time. This is possible since the lead circulates freely in the atmosphere around the world, and is then carried down by precipitation. The concentration of lead in the snow of the different ages in the Greenland Icecap has provided a record of the increasing amounts of lead, as it has with mercury. There has been a tremendous increase in the amount of lead in the Greenland Icecap since 1924, when tetraethyl lead was first used in gasoline. The amount of lead in the atmosphere of the cities of the United States that have heavy automobile traffic is approximately fifty times the amount in rural air.

At the present time, there are several major sources of lead poisoning. One of these is in the small cities or villages that lie close to the smelter plants. Lead is carried into the atmosphere where it is inhaled by all. Children are particularly susceptible. The lead can settle as a fine dust in the homes, where children are particularly sensitive to the amounts picked up on their hands and carried to their mouths, as well as to those amounts that are inhaled. Lead, like mercury, can be a waterborne, airborne, or foodborne poison.

Lead in the atmosphere was measured by a research vessel departing from San Diego, California, where the air contained from 1.5 to 2.5 milligrams of lead per cubic meter of air, to Samoa, where the air contained only 0.0003 milligrams of lead per cubic meter. The amount of lead in the atmosphere is probably hundreds of times higher, even in the relatively pure areas, than it was two centuries ago before industry reached the level it has at the present time. The level of lead in foods is quite high in many countries since plants—even leafy green vegetables—are frequently sprayed with lead arsenate. Some persons have acquired lead poisoning from game birds shot with lead pellets, wherein one or more of these pellets was ingested by the individual.

Blood levels of individuals vary widely insofar as lead is concerned. Traffic patrolmen in urban areas may have levels of lead reaching 0.3 parts per million or more in their blood, whereas individuals in mountain areas usually have 0.08 ppm, more or less, in their blood. The reaction of lead with the blood promotes fatigue which may lead to accidents, or, in some occupations, the level of lead may lead to death. Probably a million persons are at risk due to lead. Usually they are treated with chelating agents which react with lead so that it passes from the body. In some industries, unfortunately, the workers are sent back to their previous jobs and renewed exposure.

It is difficult to describe the symptoms of lead poisoning in a nontechnical manner since they may start as chronic symptoms that become severe, such as those found in painters' colic. Persons exposed to organic lead compounds such as tetraethyl lead may experience those symptoms or may have rapid onset of delirium, hallucinations which are followed by convulsions, other mental disturbances, coma, and death. Children may have encephalitis, which can result in permanent brain damage or death.

Cadmium and chromium may both be highly toxic. In the mining areas of northern Japan a number of workers have died of a degenerative bone disease produced by cadmium called *itai-itai*. Both metals can cause lesions in electroplaters, and there is a possibility that they are associated with cancer. Cadmium is found in scrap metals, rubber tires, and nickel cadmium batteries. It also has been used to plate garbage cans, causing poisoning at picnics where acid fruit drinks have been made up in new garbage cans. Airborne cadmium is related to hypertension and heart disease; and, as a result, can be of danger in areas where scrap metal containing cadmium is melted at high temperature. Certain algae may concentrate cadmium, and cause poisoning of animals and man. In poorly ventilated electroplating plants, cadmium and chromium may be respiratory poisons.

Frequently, the contents of electroplating vats are dumped illegally into the sewers, in which event these poisons are carried to the sewage disposal plants where they may destroy the selective flora of bacteria used in the activated sludge process.

Beryllium is a highly toxic element which at one time was used in the fluorescent light industry but was discontinued in 1949. Beryllium in the atmosphere is thirty times as toxic as lead, but is not associated with waterborne poisonings. It was discovered that beryllium fibers were an excellent source of power when burned in a rocket engine, but the toxicity was so great it was never used.

Asbestos is a nonmetal that produces a diffuse cancer on the lining of the chest or abdominal cavity. This can be point or non-point, depending upon the source. When inhaled into the lungs, the asbestos fibers form nodules that are covered or coated with iron. Japan had outbreaks due to the coating of rice with talc that had several million asbestos fibers per gram of talc. Asbestos was also used in the filtering of beer, wine, and soft drinks. Asbestos fibers additionally entered certain water supplies and were of danger.

Zinc is a metal which, like cadmium, is soluble and toxic under acid conditions. Both may be solubilized by soft waters or by acid drinks made in containers coated with these materials. Zinc toxicities have been more frequent where it has been associated as the radioactive isotope from the atom bomb. The zinc has been utilized by algae, then was concentrated by the tiny animal forms, zooplankton, that devour these, and eventually was concentrated to a much higher degree in the larger fishes. Actually, it is the radioactivity that is involved here and it will be discussed elsewhere.

Nitrate and nitrite are important contaminants of water in certain areas because when this water is used for infants, it can cause death due to methemaglobinemia. The nitrite ion is the most active and it attaches to and reacts with the hemoglobin of the blood, destroying its power to carry oxygen throughout the circulatory system. When nitrate is ingested, bacteria in the intestine convert it into nitrite, which is the most active form. Nitrates and nitrites may be found in any well water on farms, but they are most likely to be high in amounts in the waters on livestock farms. Here the nitrogen of fecal material of the animals and of the urea is converted to nitrate.

The only way to handle this problem sensibly is to place all of the urinary and fecal discharges into an anaerobic converter such as an Imhoff tank. By digesting the sewage in this way, sufficient amounts of methane, a flammable gas, will be produced which will supply all of the power needed for the farm and enough to sell to adjoining farms. The material which remains after digestion can then be aerated and permitted to flow into sewers, or used as a combination of mulch and fertilizer upon the farm. The water-soluble nitrogen must be handled so that it does not contribute to eutrophication, or, if used as fertilizer, must be balanced so that nitrate and nitrite do not move through the soil profile to the underlying water table.

Thermal pollution has become of greater importance in the last twenty years since it was realized that it could intensify eutrophication and in its own right be responsible for the killing of fish, or other changes in water supplies. Thermal pollution means the increase in temperature of a stream, river, or lake due to heated effluents from factories or power plants. As early as 1970, approximately 60,000 billion gallons of water were used for industrial cooling, of which 75 percent was used for the cooling of steam condensers in electric-power industries. Proportionately, generating plants fueled by nuclear energy waste 60 percent more energy than fossil fuel plants. As a result, the waste heat from a large generating plant can raise (by 10 degrees F) a river that has a flow of 3000 cubic feet per second. Since in summertime many of the rivers in industrial areas can reach a temperature of 90 degrees F, this overloading is extremely serious.

Oxygen is less soluble in warm water than in cold water, which is also true of other gases. Also, the warmer the water the greater is the metabolic activity of animals such as fish. Thus, when the activity of fish is speeded up by temperature, the less is the amount of oxygen they have available. As a result, of course, the fish probably die. With lake trout the oxygen consumption increases until a water temperature of 60 degrees F is reached; then the metabolism and oxygen consumption decrease until 77 degrees F, which is lethal. With brown trout, the oxygen consumption increases until 79 degrees F is reached, which is lethal to these fish. Even coarse fish such as carp respond in a similar fashion. They are able to live in cold waters of 33 degrees F when only 0.5 ppm of oxygen is present, but when the water is warmed to

85 degrees F they require three times this amount of oxygen because of
their increased metabolic activity.

The spawning time of fish is decreased greatly at higher temperatures, but the young are less hardy and are likely to die. Some fish can
become acclimated to warmer temperatures if the increase is made very
slowly; otherwise they will be destroyed. With most fish and shellfish,
it can be stated that they live longer and grow larger in colder waters,
but more time is involved. Experiments are in progress to determine
whether Maine-type lobsters will grow in heated waters.

The question arises: what can we do to prevent thermal pollution?
Obviously, the water must be cooled. One method is to build artificial
lakes. If sufficient land is available an artificial lake can be made. For a
1000-megawatt plant, a lake of about 2000 acres is needed. It would be
one mile wide, three miles long, and vary from approximately eighty
feet at the deep end to a few feet at the shallow end. This would be
very expensive to dig, but, once the lake was made, the water could be
recirculated and used over and over again. If the power plant is located
in an area of deep ravines or in a mountain valley, the cost of building
the lake may be modest.

Two types of cooling towers exist. One is the open or wet tower;
the other the closed or dry tower. In the open tower, the hot water runs
directly into a large cooling tower where it is cooled by evaporation.
The hot water may be sprayed from the top onto large baffle systems
and cool air, or air at ambient temperature, drawn from the bottom in
sufficient blasts to cool the hot water to near-ambient temperature. One
difficulty is that on cold days a dense fog may be formed in the area for
a great distance. If near highways, this can cause dangerous conditions
by obscuring cars and, in the winter, producing glare ice on the
highways or upon vegetation.

The closed tower, or dry tower, is similar to the radiator of an
automobile. The hot water flows through copper pipes because of the
thermal conductivity and is cooled by air blasts. This is particularly
expensive since frequently a number of miles of copper piping are
necessary for sufficient heat exchange.

It is obvious from the preceding that there are difficulties in
distinguishing non-point versus point sources of pollution. Sewage
pollution, however, is one of the most significant areas of point
pollution.

At one time, the motto of most sanitary engineers was "the solution of pollution is dilution." This motto probably originated in ancient Rome if there were any sanitary engineers who directed the *cloacae maxima* into the Tiber River, and was religiously followed in the mid-1800's in Britain and on the Continent with reference to sewage disposal. Readers who have seen sheets of toilet tissue and particles of human excrement on the bathing beaches of Britain and the Continent in the 1950's realize that sewage treatment in some areas is primitive. Since this time—although I have returned—there has been no desire to visit the bathing beaches.

The city of Los Angeles, which includes Hollywood, processes approximately 340 million gallons of domestic sewage and industrial wastes per day at the Hyperion Treatment Plant and at two smaller associated plants. Raw sewage flowing through the sewers consists of 99.9 percent water and 0.1 percent solids. Of the total amount, 75 percent is domestic and 25 percent industrial. The Los Angeles County system, which is located in a different area, is somewhat larger and uses different methods of treatment.

In the Los Angeles area a tremendous length of sewage piping is necessary to drain the city, which has approximately 450 square miles. One difficulty arises in that the system accepts the sewage of the San Fernando Valley area. The flow of water through this great length of piping is slow. As a result, the bacteria—which comprise an appreciable portion of the 0.1 percent organic matter—attack the organic materials, using up the oxygen of the water and producing a foul, stagnant odor. The amount of oxygen used at this time is not too important, but by the time sewage treatment is complete, there must be an excess of oxygen and no foul odor in the treated water.

Chemists and bacteriologists have devised a method of measuring the amount of oxygen present which is called Biological or Biochemical Oxygen Demand (BOD). For our purpose, sewage usually enters the treatment plant with a low BOD (little oxygen), but the treated effluent must leave with a high BOD (saturation or near-saturation with oxygen). If this results there will be no damage to fish or other animal life in these waters.

In the city plant, sewage enters through three large outfalls or pipes, where first all coarse debris is removed automatically by bar screens which have one-inch openings. All coarse materials, such as boards or cardboard, is ground in disintegrators to tiny particles and returned to the flow. The total waste then flows through grit tanks where sand and heavy material is removed. The remaining flow is split so that equal portions enter twelve large tanks; 300 feet in length, fifty-six feet wide, and twelve feet deep. The flow is sufficiently slow so that grease will rise to the top and heavier organic materials will settle to the bottom. Continuous chain belts with wooden cross-pieces continually scrape the sediment from the bottom and the grease from the top, feeding these into Imhoff tanks or digestors. These tanks have little or no air (anaerobic) and the bacteria present produce approximately half methane gas and half carbon dioxide.

The methane gas can be burned and enough is produced not only to power the entire sewage disposal plant, but also to furnish gas to the electrical generating plant nearby. The carbon dioxide can be purified and converted to compressed gas or dry ice.

In the digestion tanks, after a long period of time, nothing remains in the sludge but lignins or similar compounds which are inert. In Los Angeles, this residue is pumped through a pipeline which is eight miles long along the ocean bottom to the beginning of a submarine canyon. The sludge goes deep into the ocean and does not accumulate near the pipeline. Inspections are made annually by submarine. It should be noted that this does not apply to all sewage plants, for at many, raw sewage is fed directly into the ocean.

The liquid effluent, after the removal of the sludge and grease, comprises most of the sewage. In Los Angeles, this primary sewage is pumped to activated sludge tanks which are approximately the same size as the primary tanks. Here, however, compressed air is pumped to the bottom of the tanks at a rate of approximately 150,000 cubic feet per minute. The fecal organisms are overgrown by special bacteria and protozoa which form a gluey or flocculent mass that settles out in six to eight hours. During this time almost all of the organic material has been digested and the mixture is pumped to final settling tanks, where the clear water is pumped through a five-mile pipeline into the ocean. In Los Angeles, 100 million gallons a day of secondary-treated water and 250 million gallons of advanced primary-treated material flows through this pipeline.

This water from secondary treatment is sufficiently clean and clear that it can be chlorinated and used to water freeways and golf courses, or for irrigation. It can also be pumped into the underground water table to prevent the intrusion of salt water, as has been done near Sacramento; to form a reserve of fresh water after sufficient time has passed; and to raise the land level which has sunk due to overpumping as at Long Beach, California, or other western cities.

There has been no great argument with regard to the clear effluent from the settled activated sludge tanks. This material is almost completely odorless in properly run plants, and can be rendered completely odorless by what is known as tertiary treatment. One method is to filter the water through activated charcoal, at which time it is completely freed of all tastes and odors. There is a possibility that agents of disease, particularly virus particles, can be present. There are a number of ways of rendering the above water quite safe, such as the lime-alum treatment of drinking water followed by chlorination. If the water contains pesticide residues, carcinogenic compounds can be formed by overchlorination. Possibly in the future a process known as reverse osmosis, which would render the water safe, will be used more commonly. At the present time it is too expensive for the treatment of household water.

A great argument exists, particularly in the Los Angeles area, as to whether the anaerobic sludge should be permitted to be pumped through the seven-mile pipeline to the San Pedro ocean canyon where it is dissipated, or whether some other method must be used. The Los Angeles *Times* of November 11, 1978, points out that the populations living in the desert areas are unwilling to have the sludge of Los Angeles hauled to their vicinities and used as a landfill. I tend to agree with them. I realize that the ocean is not a limitless sewer and anything entering the ocean as waste should be rendered as safe as possible; the sludge, however, is relatively inert. But if the present sludge were carried to the desert areas, there are sufficient toxic minerals, phenols, and cyanide so that these would leach in the underground water table and render this permanently toxic.

On March 1, 1980, the city of Los Angeles lost its appeal to continue putting its sludge into the San Pedro submarine canyon. On March 28, 1980, it was agreed that the city, with federal financial assistance, would dry the sludge. This would be burned in special furnaces and the power produced would add to that excess now available at the Hyperion plant at El Segundo. The ash that remained would be buried in landfills.

This does not seem to be an intelligent solution, since part of the soluble toxic metal salts can leach through the sludge drying beds and infiltrate through the soil, entering the water table where the fresh water and saline waters meet. Of the remaining toxic salts, mercury, arsenic, and the others will be volatilized into the atmosphere. If they are removed with scrubbers and added to the ash, these compounds and the toxic compounds such as copper, nickel, vanadium, molybdenum, and others will eventually infiltrate into the ground water where they can create harmful results.

Even if the dried sludge were proposed to be used as fertilizer the complaints would be greater, for many of the toxic components could be taken up by plants, which could then become poisonous. The motto of every community should use the words of Shakespeare:

> If you prick us, do we not bleed? If you tickle us, do we not laugh? If you poison us, do we not die?

The situation can only increase in seriousness unless industry and the government of the United States, and actually all industries and governments of the world, cooperate. This can be done. World co-operation has almost eliminated smallpox, plague, and other diseases, and no country should be permitted to pollute the environment in which all must have pure water, pure air, and other resources that should be available.

This means that sewage sludge should not be dumped in canyons or ravines, which would intensify the problem by pollution of the underground water table. What is needed is that the toxic waste materials be concentrated, preferably at each industrial plant, and either reused or converted into materials that can be either sold or disposed of readily.

The greatest danger lies in cities located on inland waterways. A few cities such as Chicago now move their sewage sludge by barge to landfill areas, where it is hoped that toxic chemicals and disease-producing organisms will not pollute the water supply of other communities. Most, or many, cities on inland waterways dump the primary- or secondary-treated sewage and the sludge into the rivers, hoping that the water treatment plants downstream will eliminate toxic materials and agents of disease.

At times the stupidity of governments or of the judiciary can be alarming. In January 1979, eight officials of the village of Carpentersville, Illinois, were jailed because they refused to issue sewer permits for a small subdivision. They stated, and probably correctly, that the sewers were already overloaded and that if any more homes were

connected to the sewage system, raw sewage could back up into neighboring homes and also spill over into the Fox River. The Federal
District Judge held them in contempt and had them jailed for trying to
protect the community.

In March 1979, the California State Resources Control Board
issued a permit for a new sewage treatment plant to the city of San
Francisco. The new facility would permit an average of eight overflows
of raw sewage into the Pacific Ocean annually, in addition to the 114
overflows that occur at present.

Seven years ago the U.S. Navy promised that its ships would no
longer pollute San Diego Bay with raw sewage. In 1979, at least
eighteen of the 100 warships stationed in the bay had no sewage-
holding tanks; but the Navy was controlling this sewage problem at the
end of 1981, although exact figures have not been issued.

In March 1979, the International Joint Commission of U.S. and
Canada announced that pollution of the Great Lakes was again be-
coming serious. More than 2800 toxic chemicals have been found in the
Great Lakes Basin.

The Scott Paper Company pleaded no contest to ten criminal
violations of the Federal Clean Water Acts, and, according to the *Los
Angeles Times* of January 8, 1979, paid a million dollars to settle state
of Wisconsin and federal suits. The company also closed its Oconto
Falls pulp mill February 21, 1978, which was the largest single source
of pollution in the state. There are dozens or hundreds of similar cases
of contamination of the water supply.

The industries producing most water pollution in the United
States include steel mills with primary metal wastes—e.g., those from
blast furnaces, rolling mills, steel pickling, and electroplating—which
were among the factors that almost ruined Lake Erie and other areas.
Some of these plants produce cyanide, phenols, acids, alkalis, and
insoluble oils. Cyanide wastes can be destroyed with hydrogen peroxide,
which converts them to ammonia and carbonate or to ammonia and
formic acid. All of these plants need individual and specialized sewage
treatment plants of their own before wastes are discharged. In many
industries it has been proven that the waste products are sufficiently
valuable to pay for all treatment.

Paper and pulp industries produce wastes that are so high in
Biological Oxygen Demand (BOD) that fish below the areas in which
the effluent of the plants is discharged are destroyed, and the water

becomes unusable. These are complex industries, as they include treatment of wood pulp with lye, sodium sulfide or sulfite, etc. They also
include the de-inking of waste papers, the liberation of waxes and
greases, and other compounds. As a result, every pulp or paper mill
should have its own sewage plant, and the waters should be treated and
impounded until they are sufficiently pure to enter major water sources.

The petroleum industry is probably the best industry in the treatment and reuse of its waste waters. Part of this efficiency is due to the
fact that much of the water is used for cooling, so that after treatment it
can be recycled in the plant.

The food industries, textile industries, and tanneries all produce
wastes that are high in BOD, and as a result, each must have its own
sewage plant where the water is treated and oxygenated. Obviously,
each has special problems of its own, but these will not be discussed
since they are of a technical nature.

The quality of our waste waters has improved to some extent
because the government has demanded that all detergents used be biodegradable. This means that they can be broken down in the sewage
treatment plants and not remain for weeks or months as did detergents
used previously.

In certain industries where very toxic waste products that cannot
readily be rendered nonpoisonous are produced, deep well injection
has been used to dispose of these products. It has also been used for
radioactive wastes. This can be a dangerous method. Basically, it
means that a well is drilled through impervious rock below the water
table until a cavern, or water-retaining strata, is found. The wastes are
then pumped into this area. In Denver, Colorado, where this has been
done, it has resulted in slippage of rock profiles that led to a series of
earthquakes, fortunately minor to date. And underground storage of
gas under pressure has demonstrated that few strata are free of cracks
and crevices; where these exist, leakage of wastes can occur. In other
areas the underground strata or caverns may slant upwards, infiltrating
the water supply many miles away.

Decisions must be reached with regard to the highly toxic, non-
radioactive wastes from chemical plants, petrochemical industries,
pharmaceutical, and metal-plating industries. It would seem that the
one sensible means of disposal would be to convert these compounds
into useful materials.

It should be realized that normal sewage from either homes or hospitals in which persons are treated with antibiotics, or wastes from pharmaceutical industries, enter municipal sewage systems and can result in the production of disease-producing bacteria which are resistant to these antibiotics. Resistance from relatively nonpathogenic bacteria can be transferred to organisms that are highly virulent and nonresistant to drugs.

The greatest danger of sewage-polluted waters lies in the transfer of disease. Mass epidemics of cholera, typhoid fever, and other diseases occurred before the danger of sewage pollution was realized. In London in 1854, an outbreak of cholera occurred among people who obtained their water from a pump on Broad Street. Dr. Snow, an eminent physician, was asked to investigate. He recognized that this well lay adjacent to a privy in a yard. One of the members of the household had cholera, and the fecal material leaked through the broken masonry of the well. Snow's answer can be reduced to: "Gentlemen, in order to stop the cholera epidemic, remove the handle from the Broad Street pump."

In 1892 a group of Russian lumber ships anchored in Hamburg harbor in Germany. Certain members of the crew had cholera and their fecal matter, as well as all other sewage, was discharged into the harbor. The city of Altona, which lay below Hamburg, had become tired of drinking the sewage-polluted water of the larger city above and had installed a water filtration system. As a result, Altona experienced almost no cholera, whereas Hamburg, which used unfiltered water, had a major outbreak. As a result of this outbreak, almost all major cities installed water filtration plants. As Thomas Mann stated, "All interest in disease and death is only another expression of interest in life."

The most frequent and dangerous type of water pollution is derived from fecal and urinary pollution of water. From the time of the fall of Rome, which had uncontaminated lakes and aqueducts, until the twentieth century, most city water supplies have not been free of disease-producing organisms. To mention only a few, some of the larger forms of animal life such as hookworms, schistosomes, and intestinal roundworms can invade through the skin or by mouth. Many of the laborers who must work barelegged in irrigated areas of Egypt or rice fields of the Orient develop a debilitating disease called schistosomiasis. Other larger animal forms such as flatworms and flukes may be derived from raw fish or use of polluted waters.

A smaller animal parasite causes amebic dysentery, which is common in Mexico and much of Latin America, Africa, and the Orient. It is a severe disease which must be treated promptly and thoroughly. It can be obtained by using polluted water directly or by the use of salad vegetables, and even fruit, handled by persons with fecally polluted hands. The safe rule applied to foods in these areas is heat it, peel it, or don't eat it.

Certain bacterial infections are spread by sewage-polluted waters or contaminated foods. These include typhoid fever, bacillary dysentery, cholera, leptospirosis, enteropathogenic *Escherichia coli*, and a host of other diseases.

Viruses such as infectious hepatitis and poliomyelitis can be present in the waters from sewage pollution and can be contracted directly from the water, or by eating oysters or other raw shellfish which have become contaminated. It is unsafe from the standpoint of any of the diseases mentioned to swim in fecally polluted waters or to use foods in which these waters may have come in contact.

It is impossible for the health officers to test for each organism of disease so they test, ordinarily, for fecal *Escherichia coli*, an organism always present in the intestinal tract in large numbers. If these organisms are present it is assumed that disease-producing organisms might be present and the water supply should be treated.

Approximately half of the world's population, or two billion men, women, and children, do not have a safe and adequate water supply. In developing countries forty of every hundred children die by the age of five years primarily from waterborne diseases such as typhoid, cholera, dysentery, and intestinal parasites. The World Health Organization estimates that the source of 80 percent of all diseases in the world is waterborne and that infant mortality would decrease by half worldwide if all had access to pure and safe water supplies. Portions of southern United States and Appalachia also suffer from this deficiency. As Robert Traill Spence Lowell said, "Christ walks on the black water."

For the traveler in Africa, Latin America, and the Orient, it is advisable to carry a kit of either chlorine or iodine tablets and to drop the suggested amount into all water used for drinking, brushing the teeth, and washing vegetables. The chlorine or iodine tablets should be present in the water for fifteen or twenty minutes, at which time a tablet of thiosulfate can be added that will convert the chlorine into salt. The level is so low it is tasteless.

A new danger has arisen relative to the transfer of disease that the health officer cannot measure. Wastes from lavatories of airliners may be spreading cholera and other diseases around the world. Cholera has been discovered in areas such as the Dordogne area of southwest France and the Essons district in the north in the past year or two. These areas are all beneath the regular flights from India to the West. Most airliners hold toilet refuse in tanks for discharge at airports, but waste water from the lavatories is discharged directly into the air. Passengers who are carriers of cholera or other diseases can disseminate these in the areas of the flight of the plane. The organisms will survive a release into freezing high atmospheric air and fall to earth in a fully virulent condition to cause illness.

The Air
We
Breathe

Today, as I am writing in my den on Mount Washington in Los Angeles, the San Gabriel Mountain Range seems almost close enough to touch. As I turn my head back for a moment to the northwest, the Santa Monica Range is clearly visible, and to the east of the San Gabriels the San Bernardino Range is the most majestic of all. The peaks of Mount Waterman, San Gorgonio, and Old Baldy are snow-covered and the tops appear to be mounds of vanilla ice cream piled upon inverted gray cones. The air is crystal clear and the television towers of Mount Wilson seem to be only three or four miles away. From the vinyl-covered sundeck, the Santa Anita Range is visible and these all enclose the Los Angeles Basin in a vast semicircle.

It rained last night and all the day before, and at 4000 feet or higher there has been a tremendous snowfall. The television pictures show the highways that are blocked, with schools and churches turned into refuges along the mountain passes. From here, however, the air is so clear that the individual homes that comprise Eagle Rock, South Pasadena, and parts of Pasadena are clearly distinct, the white or pastel colors shining like jewels. I believe, like Wordsworth, who could wish for, "an ambient ether, a diviner air."

Life today is enjoyed to the fullest. Cars move east or west on the Ventura Freeway with drivers who probably do not know why they wish to leave this area to go to a different place on such a beautiful day.

They realize, perhaps, that within a week the area will be covered by photochemical smog that will blot out the base of each mountain and probably obliterate a clear view of the Ventura Freeway and the beautiful pastel-colored homes. From Mount Washington, the snow caps of Old Baldy and Mount Waterman will be visible, but the smaller peaks will have disappeared.

At our home conditions will be relatively comfortable, for there are few days only in which the sting and smart of smog will burn the eyes and irritate the nose. When we wish to drive to the university from which I recently retired, however, it is necessary to drive down smog gulch, the Pasadena Freeway, cross the four-level stack of freeways at midtown, and remain in the smog until we return home.

Days away from home are quite few for the retired hill-dwellers. It is preferable to stay amongst the bougainvilleas and natal plums, and the beautiful Monterey pine and olive trees that shade the front and back patios. I still use my office at the university two days a week and on smoggy days try to imagine that the air conditioner is of some service. If one moistens the filter it does help to some extent. After our writing and use of the library is complete, and we have had lunch at the Faculty Club with colleagues, we drive back to our refuge as promptly as possible.

The difficulty with Los Angeles smog and conditions encountered other places in the world is that the semicircle of mountains can retain the noxious compounds formed. Photochemical smog is different from the materials and maintenance effluent found in most parts of the world in that this smog requires light (*photo*) and certain chemicals to form. Once it has been produced it may be maintained in the Los Angeles Basin by the encircling mountains and a peculiar weather condition known as an inversion layer.

This smog is completely different from the accumulations of sulfur dioxide, fog, fluorine, large soot particles, and other conditions in most cities. Actually, Los Angeles is one of the cleanest major cities in the world, with relatively fewer dust or other particulates per cubic meter of air. Photochemical smog can be formed in other places in the world, but there are none with the intensity of sunlight to form the smog and the encircling mountains to retain it. Los Angeles is the home of photochemical smog and the place in which it has been studied most intensively.

Historically, Juan Cabrillo noted on October 8, 1542, that he could see the mountain tops in the distance but not their bases from San Pedro Bay. The smoke of Indian campfires obscured the lower levels. For this reason he called the area "La Bahie de los Fumos," or the "Bay of Smokes." Cabrillo had arrived during the time of an inversion layer of air. The phenomenon that Cabrillo noted is magnified a million times by the automobiles and industry of the area in this year 1982.

Photochemical smog is produced when approximately a million workers press the starters of their passenger automobiles between 7 A.M. and 8 A.M. at least five days a week. There are two million more passenger automobiles that will be started later by the executives who do not have to be at the office early, or by the housewives who must take children to school or run miscellaneous errands. There are also innumerable trucks and buses powered by diesel fuel operating at the same time.

As the engines start, the tail pipes belch out the fumes of partially burned and unburned hydrocarbons which are the constituents of gasoline, and a white-to-bluish grey cloud is emitted. If the engines are highly tuned and run "hot," nitrogen dioxide, which produces a brown toxic gas, is emitted in great abundance. In addition, poisonous carbon monoxide, carbon dioxide, and residuals of lead are blown out if unleaded gasoline is not used.

The nitrogen dioxide decomposes to form nitrous oxide and atomic oxygen, which immediately reacts with a molecule of ordinary atmospheric oxygen to produce ozone, a toxic gas. Ozone reacts with the hydrocarbon molecules to produce a whole chain of compounds—which will not be mentioned here—in the presence of sunlight. As a result, the reaction becomes more accelerated until approximately three or four o'clock in the afternoon. At this time the sun rays are losing part of their intensity and the reactions slow down and finally stop. Since ozone is a readily measurable reactive factor, photochemical smog is measured in parts per million of ozone for which quantitative tests exist.

The photochemical smog accumulates in the Los Angeles Basin because of inversion of the atmosphere. In most areas of the world, the air nearest the ground is the warmest and it becomes colder and colder in a straight line the closer it approaches the stratosphere or upper air. In Los Angeles the conditions are different more than 200 days of the year. A warm air front moves into the basin area at approximately 2000 to 3000 feet, overlaying the colder air below. This colder air from the Pacific Ocean slowly fills the bottom. Above the warm air front is the usual cold air, and, as a result, an air sandwich is formed in which a warm air layer lies between a lower and an upper cold layer. The warm air layer holds down the photochemical smog, which moves slowly backwards and forwards across the basin with the gentle breezes.

Contrary to most textbooks, at approximately four o'clock in the afternoon the breeze from the Pacific Ocean blows the smog from the west or southwest towards the northeast. At this time the photochemical smog is piled against Pasadena, Altadena, and the mountains. In a long period of smoggy days this becomes so dense that it crosses through the mountain passes and becomes a nuisance in the high desert and the low desert areas. There are days upon days when it rests as a gray-brown or yellow-brown blanket over the whole Los Angeles Basin during the times when the inversion layer acts as a cover, concealing a seething pot of poisonous brew.

Individuals flying from the east are aghast to look down upon, and then descend into, this irritating area. The conditions will remain like that until there is a climatic change, and one may not glimpse the beauties sung by Thomas Gray:

> The meanest floweret of the vale,
> The simplest note that swells the gale,
> The common sun, the air, the skies
> To him are opening paradise.

If the winds blow from the east across the Mojave Desert, the smog is blown toward the Pacific and covers Catalina Island so that it is no longer visible. It is during these times of the Santa Ana "devil winds" that the smog reaches my home and we try to prevent the worst effects by using a five-ton central air-conditioner with appropriate filters.

Of the total of about 2600 tons of hydrocarbons that enter the air of Los Angeles County each day, 1800 tons are from automobiles, 500

tons from organic solvent usage, and 240 tons from the petroleum industries. Other sources contribute the remainder. If the smog becomes
sufficiently bad, a "smog alert" is called. Athletics at school are discontinued as it has been shown that lung damage can occur. Certain
industries must decrease or cease their activities, and supposedly, automobiles are to be driven as little as possible. Trucks and buses still belch
out their diesel fumes. Since compulsory school busing has come into
Los Angeles, it will have added somewhat to the percentage of smog.
Buses now travel from twenty minutes to over an hour to achieve inte-
grated schools. When only a few children are transported fairly long
distances, one must weigh the resulting damage on the scales with the
sociological benefit being sought.

Near the latter part of October 1978, it was reported by Tom
Quinn, Chairman of the Air Resources Board, that infants born in the
smoggiest areas of Los Angeles weigh eleven ounces less, on the average,
than children born in less-polluted areas. Infant mortality in the
smoggiest areas of Los Angeles may be 17 percent higher than in other
areas of the basin. Both of these points are probably factual, as it was
established a decade ago that babies born to mothers who smoked
cigarettes weighed, in general, less than those of nonsmoking mothers.
It has been known for a long time that smoking tends to decrease the
transportation of oxygen across the placenta, thus robbing the developing fetus of needed nourishment. Air pollution in the smoggiest
areas can easily be equivalent to the mother smoking a package of
cigarettes a day.

It has been mentioned that photochemical smog is both an eye and
nasal irritant. In addition, emphysema is probably the cause of death
increasing most rapidly in the United States, or at least in the smoggy
sections of the country. In the last two decades there has been a tenfold
to twentyfold increase in deaths due to emphysema. Those individuals
dying of lung cancer are fortunate in comparison to those dying of
emphysema, since emphysema is a condition that may last for years
with its victims feeling that they are strangling almost every moment.
Lung cancer is fast-moving and not as painful. The latter as well as the
former are correlated with smoking and with photochemical smog;
where both are present, the effect is accelerated.

The individual with lung cancer will die more rapidly than the
patient with emphysema. The only salvation for the latter is to remove

the patient from smog-filled air and prevent him or her from smoking. One of the most frustrating experiences for an internist is to have saved a person from death by emphysema by enclosing him in an oxygen tent, only later to hear the first words, "Doctor, can I have a cigarette now?"

Lung tissue is highly absorptive and very sensitive to air pollutants of any kind. Nitrogen dioxide of the smog or other toxic materials produce patchy congestions throughout the lung, which cause edema. Drugs are of little or no assistance as the high concentration of pollutants destroy the lung surface areas. When the individual has been exposed to ozone or nitrogen dioxide at a significant level, he or she is also more susceptible to lung infections. The individual living near busy roadways or highways has a risk of developing lung cancer approximately nine times that of persons in other areas, due to the polycyclic aromatic hydrocarbons.

It should be mentioned that emphysema not only is caused by photochemical smog and its constituents, but also can be caused by industrial dusts such as those of cotton, asbestos, or other small fiber-like materials, even though many of these diseases have been given specific names. We will refer to specific industrial diseases shortly.

Photochemical smog may have synergistic effects with other diseases, in that individuals whose bronchial tract and lungs have been damaged by smog may fall victim to other respiratory diseases, lung cancer, or even heart anomalies. A disease known as Tokyo-Yokohama respiratory disease is quite similar in symptoms to those experienced in the Los Angeles Basin during the smoggy season; it is primarily asthmatic in symptoms. The symptoms are particularly severe amongst smokers.

Associated with the photochemical smog produced in Los Angeles, approximately 10,000 tons of carbon monoxide per day are emitted from the three to three and one-third million automobiles. This will establish a general level of fifteen to twenty parts per million of this deadly gas and, at certain times, such as rush hours on the freeways or in the tunnels, this can reach 100 parts per million. This is quite a toxic amount if the individual is exposed for any length of time. These concentrations can cause dizziness and headaches. Carbon monoxide does not enter the smog formation reactions. Because of its toxicity, however, it can damage persons continually exposed, such as policemen in the downtown areas of the city. In the United States, more than sixty million tons of carbon monoxide are emitted into the air each year. Carbon monoxide greatly increases the risk of angina attacks.

The author has the latest reports on photochemical smog from all parts of the Los Angeles Basin, but these statistics are meaningless unless one is a specialist in the area. Despite the damage of photochemical smog, the Environmental Protection Agency agreed in the early part of 1979 to lower its standards by 50 percent as a fuel-saving device. This means that, at present, 0.12 micrograms of ozone per cubic meter of air are permitted instead of the previous standard of 0.08 micrograms. Needless to say, the Los Angeles Basin has not been able to attain either consistently.

It is necessary to continue our war on photochemical smog. It has been mentioned that nitrogen oxides, hydrocarbons, and carbon monoxide produced from the automobile and stationary sources, which injure humans and animals, also damage plants. Two of the components, ozone and peroxacetyl nitrate (PAN), are extremely toxic and the latter can damage plants at a concentration as low as five parts per billion. Since this compound is relatively stable, its effects are more far-reaching than those of the ozone-olefin compounds that last only a few moments. Nitrogen dioxide is also highly toxic for certain plants and is injurious at levels of about three parts per million. It is of interest that whereas carbon monoxide is more concentrated in the downtown areas, ozone is somewhat more concentrated in the suburbs.

Photochemical smog is not only responsible for hidden injury in plants due to decreased growth rate and increased susceptibility to plant pathogens, but also for reduced water uptake that decreases the fresh weight of fruits and vegetables. In the above, due to plant injury, many plants—whether grown in the flower garden, like roses, or fruit such as citrus trees—are damaged to the extent that aphids, mealybugs, and other parasites can invade and damage them more readily. For the home gardener interested in cucumbers, cantaloupes, table beets, parsley, and many other vegetables, the direct damage due to smog and the indirect damage due to the parasites attacking the weakened plants make it almost impossible to grow these. In addition, these same crops cannot be grown profitably in many areas of the Los Angeles Basin, due to the same factors. In June 1981, Walter Beck, a North Carolina research worker, told a House Panel in Washington, D.C., that air pollutants damage or destroy 2 percent of the nation's crops, worth at least one billion dollars.

It is not our intent to develop the injury of photochemical smog on plants to any great degree as the subject is much too complex. It should be mentioned that ozone causes a partial closure of the stomata of leaves, which are the small orifices in the epidermis, or outside. As a result, photosynthesis (the ability to use sunlight as energy to produce certain organic compounds) is reduced, and transpiration (the ability of plants to give off water) is also diminished. Ozone also causes bleached or light areas, which are clusters of dead cells on leaves. Fully expanded, mature leaves are readily injured and the disease of white pine known as emergence tip burn can be caused by 0.05 parts per million or less. A very severe form of leaf mottle on *Pinus ponderosa* and other conifers has been found near Lake Arrowhead, which is high in the mountain area. Young trees are also stunted in growth.

In citrus groves the trees suffer from both a lack of water and photosynthesis when compared to plants shielded from photochemical smog. Both of these are probably related to the effect of ozone on the stoma of the leaves. When nitrogen dioxide was added to reduce ozone, the total effect of the nitrogen dioxide and reduced ozone was as damaging as that of the ozone alone.

When Zinfandel grapes were grown in normal field conditions and under protected conditions, the photochemical smog reduced the yield of grapes and produced a severe leaf stipple, showing that the ozone was interfering with the ability of the leaf to produce chlorophyll, the green pigment necessary for photosynthesis. Absent would be the cleanliness of nature hypothesized by Swinburne:

> Ah, yet would God the flesh of mine might be
> Where air might wash and long leaves cover me;
> Where tides of grass break into foam of flowers;
> Or where the winds' feet shine along the sea.

Few purchasers of now glamorized "resting places" in this area can look forward to the rest visualized by Swinburne in idealized surroundings, since cemetery plots are now showing the erosion of photochemical smog. Yet, our dwellers in the City of Angels seem largely indifferent, as was John Donne, to the faster encroachment of the Grim Reaper:

Death be not proud, though some have called thee
Mighty and dreadful, for thou art not so,
For those whom thou think'st thou dost overthrow,
Die not, poor death, nor yet canst thou kill me.

Los Angeles, as a city at least, recognizes its problem and although it fumbles to the right or to the wrong, the intent to solve the problem is present. An article in the Los Angeles *Times* of January 24, 1979, states that the damage to crops is reported to be $55 million each year. This figure is low. Also, if public health costs were added, the total would approximate $1.5 billion each year when the problems of emphysema, lung cancer, and increased susceptibility to respiratory diseases are taken into account. Dr. Mohammad Mustafa of UCLA stated in September 1979 that while direct links to cancer by air pollution have only been strongly suggested, his experiments with animals indicate that they undergo premature aging and death.

Unfortunately, this area is devoid of a rapid transit system and one should be established. It has been proven that if one lane of each freeway were allocated to high-speed buses, and almost all private automobiles eliminated from the city's center, the problem would decrease markedly.

In addition to the hydrocarbons, of the other important gases 10,000 tons of carbon monoxide enter the atmosphere of Los Angeles County, almost all of which are derived from gasoline-powered motor vehicles. At least 900 tons of nitrogen oxide are also given off into the air each day, of which two-thirds are derived from the automobile and almost one-third from other sources of combustion. In addition, 180 tons of sulfur dioxide enter the atmosphere, of which less than one-fourth is from the automobile, the greatest single source being the chemical industries. The ratio of these figures for Los Angeles County is not badly out of proportion with those of the United States as a whole, in which the sum total of these gaseous pollutants is about 150 million tons per year. Aircraft, of course, contribute a noticeable amount of those ingredients causing photochemical smog. But even where two or three planes are landing or taking off per minute, the speed is such that the contribution is of short duration.

It must be mentioned that the ozone of smog results in the cracking of rubber, which can be noted on the sidewalls of most automobile tires in the Los Angeles area. Last and not least, the author mentioned the view from his vinyl-covered sun deck. The view is magnificent on smog-free days, but the atmosphere is really tough on the vinyl. It has been necessary to paint this deck with at least one coat of vinyl paint each year to prevent the checking and ruining of the surface.

Of the pollutant gases not discussed as yet, sulfur dioxide is the most widely spread throughout the world. It is present in both coal and oil, the conventional fossil fuels. Coal may contain much more than 3 percent sulfur. What has been termed "sea coal" in Britain contains much more, and when this was burned it gave off an obnoxious odor of sulfur dioxide and black particulates of soot.

In 1306 or before—possibly in both 1273 and 1306—a proclamation was passed by Parliament that only wood was to be burned at those times when Parliament was in session. Approximately a year later, an individual was tortured, and presumably executed, at the beginning of the reign of Edward I for violation of this edict. This regulation was revived and enforced during the time of Elizabeth I. During the time of Charles II, John Evelyn was actively writing against the peril of air pollution and in 1661 addressed a pamphlet *Famifugium* to His Majesty, in which he described this evil practice of burning sea coal. He made the very sensible suggestion that factories using this fuel be moved further down the Thames Valley and that a "Green Belt" of trees and flowers be planted to protect the city. Approximately 250 years later, this suggestion was finally implemented.

(The effects of sulfur dioxide are particularly bad, as this gas combines with moisture of the atmosphere to produce sulfuric acid. Marble statues in heavily industrialized cities where high-sulfur coal has been used for long periods of time are badly etched. Splendid statuary and other works of art in Italy are paying a price to industrial pollution; the magnificent remnants of Pompeii are being rapidly eroded, as is the Parthenon in Athens. The same bad effects of this gas on lung tissue occur and were recognized in the seventeenth century by the famous physician Thomas Sydenham.)

• London was known as a dirty, grimy city in which the air was filled with soot particulates. These, mixed with the sulfur dioxide and fog, exposed those persons living in these crowded areas to repeated attacks of bronchitis and pneumonia, if the first invasion of the latter disease was not fatal. Chimneys were kept relatively clean by chimney sweeps, who were prone to develop cancer of the scrotum from the carcinogens deposited by the coal smoke. Undoubtedly, many instances of cancer of the lung also occurred, and in those years tuberculosis or "consumption" was quite common, possibly among the chimney sweeps also.

• In 1930 the Meuse Valley in Belgium was covered by a stagnant air mass. This valley was the center of blast furnaces, coke ovens, glass factories, power plants, and steel mills; and on the first week of December a classical inversion occurred, trapping the inhabitants in the toxic gases. Over a thousand persons became ill and sixty died as a result of the smog, which was not photochemical.

• Later in October 1948, the town of Donora, a small mill town in Pennsylvania on the Monongahela River, had a similar episode. The town contained a zinc-reducing plant and a steel mill. For three days this town, which is surrounded by low hills, suffered an inversion in which the toxic gases caused at least thirty-six deaths, and illness in almost half of the population. This was attributed to sulfur dioxide and the fine particulates present in the air.

• On December 4, 1952, London was under siege in the same way. A slow-moving high-pressure system spread slowly southeast across the British Isles. Fog formed that evening and the next morning, but at this time tons of smoke from the millions of domestic chimneys and the power plants mixed with the fog, all adding their particulates plus sulfur dioxide. The cars, buses, and trucks added their fumes and the city was held in the grip of a mixed fog turning from yellow to brown and black. Within twelve hours of its onset people began to die from its effects. The exact number that died as a result is not known, but it is believed that this siege resulted in 4000 deaths. London experienced a second episode in 1956 that caused a thousand deaths, and a third episode in 1962 in which 700 died.

The deaths that occurred in the outbreaks above were all acute cases. The chronic situations that occur in polluted atmospheres include chronic bronchitis, bronchial asthma, emphysema, and lung cancer. Any of the damaging air pollutants can affect the tiny hairlike cilia of the respiratory tract. These have a peristaltic, wavelike motion that continues to eject particulates that enter. If the atmosphere contains significant amounts of ozone, sulfur dioxide, nitrogen peroxide, or fluorine, the particulates in the air produce various degrees of damage. In chronic bronchitis, due to the failure of cilia to act, mucus accumulates and a chronic cough develops. The opening of the bronchial system is obstructed by the mucus and a shortness of breath develops.

Bronchial asthma occurs when the bronchial membranes are allergic to foreign proteins or to particulates. There is difficulty in expelling air from the lungs, and wheezing and shortness of breath may develop.

Emphysema results when an individual cannot expel most of the air from the alveoli, the tiny air sacs of the lung. When new air is inhaled, the alveoli may rupture, reducing the number of alveoli and destroying the capillaries that carry oxygen. This results in a slow starvation for oxygen throughout the entire body and chronic shortness of breath—one of the longest and most painful methods resulting in death.

Lung cancer is caused by any number of carcinogens, such as benzopyrene in cigarette smoke and numerous others. Those interested can find these listed in any text of internal medicine. Particulates of many kinds entering the lungs can cause special types of cancer.

The damage of acid rain and snow due to the action of sulfur dioxide from the belching smokestacks of power plants was realized in 1980, and President Carter asked Congress to approve a $100 million program to assess the acid rain problem. Not only does the acid falling from the skies destroy the trout, bass, and other game fish, but it also has a deleterious effect upon the soil. The acid can destroy the finish of automobiles and erode the stone statues built by our predecessors— statues that withstood the thousands of years from ancient Egypt or Rome. More importantly, perhaps, to our personal concern, it damages the human lungs.

Studies in the United States are just beginning to demonstrate the damage inflicted by the fifty million tons of sulfur dioxide spewed into the atmosphere from the stacks of smelters, refineries, homes, and automobiles. The effect of acid rain in Japan and Canada is just being studied also.

In 1981, Canada recognized the effect of air pollution from the United States relative to both acid rain and acid snow. A study of the danger of acid rain was promised by officials of the Argonne National Laboratory; the most polluted air of Chicago was mixed with rain water and measurements of the acidity followed. Had this been a proposal of the previous administration, we would have said "peanuts." Under the present administration, there is a definite policy to evaluate health and other damages of an industry against the economic gain.

If one realizes that a 1000-megawatt power plant that burns coal having 10 percent ash, 1.5 percent sulfur, and 1.5 percent nitrogen will produce 900 tons of carbon dioxide, twelve tons of sulfur compounds, over five tons of nitrogen oxide, three to five tons of particulates, and thirty tons of ash per hour, it is not difficult to realize why the establishment of coal-fired power plants is fought. It is necessary to install scrubbers in some plants to eliminate air pollution. One standard of the Environmental Protection Agency that seems rather ridiculous is that it requires all new coal-fired electric generators to remove 70 to 90 percent of the potential sulfur oxide emissions in any coal they use. It is unfair to require utilities using low-sulfur coal to install multimillion dollar scrubbers. The act was intended for high-sulfur coals and should be limited to this.

Although there is no relation to air pollution, the injury and death rates involved from mining, transportation, and plant accidents will be as great as that in the nuclear power plants.

The atmospheric pollution by fluorine gas has not been taken up until this time because, although it is a poison, it does not produce smog in the classical sense. Although fluorine would be a deadly poison to man if sufficient exposure were possible, it is an element that is essential for the hardness of the enamel and dentin of the tooth when present in minute amounts.

Fluorine's toxic qualities in the atmosphere are particularly noted within some miles of plants converting rock phosphate, an insoluble fluorapatite, to the more soluble rock phosphate, by means of heat. The fluoride enters the atmosphere and can be carried some miles from the plant if there is a usual prevailing wind. Injury to the leaves of the plants occur and in some plants there are purple bands and areas at the edge of the leaf where chlorophyll has been destroyed. Within certain distances the plants may be destroyed. Farther away, the plants will accumulate fluoride, and, if this is in hay crops such as alfalfa or clover, animals eating the hay will become ill and can—but usually don't—die. The symptoms are usually those of lameness and stiffness of the joints. These may be puffy in appearance. If the animal is killed and autopsied, it will be noted that in severe cases the bones are spongy in appearance (osteoporosis) and may appear chalky white. The easiest method of diagnosis is by the degree of mottling, or chalky appearance of the enamel of the incisor teeth in comparison with the molar teeth.

In manufacturing plants producing rock phosphate from fluorapatite, techniques should be used to prevent the escape of the fluorine.

Lead is one of the most dangerous atmospheric pollutants and this may or may not be combined with sulfur dioxide. Lead poisoning is not infrequent in children living in ghetto areas, where they ingest relatively large particles of lead paints that have fallen from the walls or ceilings of homes painted with these toxic materials. Also, at one time lead poisoning was commonly encountered in house painters. These workers frequently would get lead on cigarettes they removed from a package or onto food materials they ate without adequately cleaning their hands. Neither of these classify as air pollution and will not be discussed here, although the poisoning of slum children is a serious problem for public health authorities.

The type of lead poisoning that is important in the air we breathe is that similar to the lead poisoning encountered in the mining town of Kellogg, Idaho, where 98 percent of the blood samples of the 175 children examined showed dangerous lead concentration, according to Dr. James Bax, Chief of the Idaho Health Agency, in September 1974. Twenty-one percent of the children showed more than eighty micrograms of lead per 100 milliliters of blood, which is unequivocal lead poisoning. This lead arose from the smelter in the town. Of the blood samples examined, only two were under the "serious" level. The company, of course, released a statement that if the results were accurate, the company had complete confidence in the community doctors to effectively treat any lead poisoning encountered. Nothing was said about restricting lead emissions from the plant. In 1981 Dr. Thomas Kurt stated that lead contamination from three South Dallas smelting plants caused serious illness and brain damage among plant workers and area residents. The lead levels in a twenty-five-mile area surrounding the plant were found to be ten times higher than levels recognized as being harmful to humans.

In an article of November 3, 1978, concerning a company in Beech Grove, Indiana, it was mentioned that workers in lead smelters may be sterile. (On a national basis, possibly as many as one million workers with exposure to lead may be involved.) Normally, the treatment for persons exposed to excess lead is in the administration of a chelating agent, a chemical that will inactivate the lead and permit removal from the body, during which time the worker is removed from all contact with lead. In one situation rather well documented, the company physician seems to have given the injections of the chelating agent and sent the employees back to their previous positions. One worker had been given seventy intravenous injections of the chelating agent and sent back to the same exposure that had poisoned him. A toxicologist stated that "this was a criminal procedure."

Every individual in the Los Angeles Basin area, and most others, inhales lead particulates from the use of leaded gasolines. The lead tetraethyl is spewed into the atmosphere, and when traffic is bumper to bumper in morning or evening, or when a traffic tie-up occurs, more lead is inhaled. In addition, when crops are grown along the roadside they contain much more lead than crops grown two or three hundred feet farther from the roadway. Crops in fields near busy avenues that have four-way stop signs or traffic control signals will accumulate more lead than the same crops along the open roadway.

Fortunately for man, he frequently washes foods such as tomatoes, and much of the residue is deposited on nonedible areas of other foods, such as corn husks and stalks. Even under these circumstances, it has been estimated that the average human inhales approximately fifteen milligrams of lead per year, receives five milligrams of lead in his drinking water, and ingests 100 milligrams per year from his food materials. The author knows of no accurate measurements in the Los Angeles Basin area, but is certain the local amounts would be higher.

There are many other toxic elements that can be associated with atmospheric pollution under specific conditions. Black lung has been known to be a severe disease of coal miners that can result in illness or death. Although in recent years some of the more progressive mines have sought to remove a portion of the danger by improved ventilation, it is only by the use of face masks in certain of the jobs that the tiny particulates can be kept from entering the respiratory system. This disease is probably the least dangerous of those produced by particles or "particulates" in the air. The filter system of the nose almost completely removes particles that are larger than 1/2000 inch, or approximately ten microns. Particles that range from two to ten microns usually settle on the walls of the trachea and only the very small particles that are one micron (1/25,400 inch) reach the alveoli of the lung. The lung has other protective mechanisms, but the shape and composition of the fibers are very important. Persons living in the city, where a single idling automobile can emit 100 billion particulates per second that measure approximately 0.1 micron, are also likely to show darkened lungs.

The most serious of the pulmonary diseases are caused by silica (silicosis), asbestos (mesothelioma), and beryllium, each of which may produce tough, fibrous tissue causing serious diseases.

Probably the greatest charge of carelessness or of criminal carelessness can be brought against the asbestos industry. Mesothelioma has been known since approximately 1927. Although the seriousness of the disease was known by executives of the industry, reports were suppressed so that the workers would be unaware of their danger. It is quite possible that of the four million workers who have had heavy exposure to asbestos, 1.6 million may die of mesothelioma. In addition, from the end of World War II until 1973, schoolrooms were sprayed with asbestos insulating material. Are these areas now crumbling to fiber-size particles that will endanger the children involved? The first lawsuit against Johns-Manville and Raybestos-Manhattan, brought by a shipyard worker who developed asbestosis, went to the jury in mid-May 1980. A thousand more lawsuits are expected. Johns-Manville, in turn, has filed suit against the eighteen insurers that provide coverage insurance against such losses.

In 1981 it is known that approximately 35 percent of household contacts, the wives and children of workers in asbestos industries, show evidence of mesothelioma or lung cancer.

Cotton dust, byssinosis, or brown lung disease has killed or injured thousands with classical symptoms of coughing, shortness of breath, and tightness of the chest. Over 80,000 of the half-million workers probably have the disease and 6000 are severely disabled. The Occupational Safety and Health Agency (OSHA) had fought for years to clean up the conditions in the mills.

During the political campaigns of 1980, the successful presidential candidate criticized OSHA severely. As a result, in 1981 the Justice Department asked the Supreme Court to throw out earlier rulings on OSHA standards, including the reduction the agency had made which lowered the standards for lead from 200 micrograms to fifty micrograms per cubic meter, demanding that OSHA conduct a "cost-benefit analysis."

Fortunately, the Supreme Court ruled against the president on byssinosis, the lowering of strip-mining control laws, and certain others. It has permitted the steel industry a "stretch-out" period of three years to meet the air pollution standards.

One type of pollutant, the effects of which may be classified as air pollution, has not been mentioned until the present: the use of urethane as an insulator in homes and commercial buildings, as well as in seat cushions of automobiles, aircraft, and home furnishings. Urethane and polyurethane are plastics that almost have the flammability of gasoline. Aircraft cabin fires can reach a stage of explosive burning known as "flashover" in as little as two or three minutes. Large buildings are still being insulated with these materials, and there is no doubt that a towering inferno can occur readily by shorts in electrical insulation or from other causes.

On March 22, 1975, a worker at the Tennessee Valley Authority's nuclear power plant in Alabama, checking for air leaks around the cables of an operating 1000-megawatt reactor, held a candle near the cables. The polyurethane ignited instantly and the worker could not put out the fire. The fire smoldered and flamed for seven hours while authorities sought to extinguish it. Smoke poured into the reactor control room and hundreds of cables shorted out, which rendered the pumps, valves, and backup cooling systems inoperable. It was by good fortune and against the advice of the TVA that the Athens fire department put out the fire, which could have had tragic results of tremendous impact. There have been automobile and school bus fires in which lives were endangered or lost due to the use of these materials, when safer materials exist. In the latter part of 1979 the National Academy of Sciences warned that urethane foam, a flammable plastic, should be barred from automobiles and all public transportation.

Urethane foam, air pollution, and energy conservation can be coupled together, as may other insulating agents. An energy-efficient home completed in Maryland with triple-glazed windows, double insulation, and polyurethane caulking showed high levels of humidity, green mold odors, and a hundredfold accumulation of radioactive radon gas. In other energy-efficient buildings where urea-formaldehyde foam is used as an insulator, if the foam has not been cured properly, formaldehyde gas is released into the air of the room, causing visual impairment, difficulties in breathing, and chronic nausea.

A home that is too energy efficient may turn the rooms into gas chambers, trapping build-up of vinyl acetate copolymer resins from hair spray, fumes from cleaning fluids, morpholine from furniture polish, smoke from cigarettes, and gas from appliances. The normal home has approximately one complete air exchange every two to three hours. A home that is overly energy efficient may have only one air exchange every twenty-four hours. This can be dangerous! In our interest in conservation, we are making our homes indoor smog traps.

Normally, one does not intentionally pollute the atmosphere. In 1950, however, a Navy ship blanketed San Francisco and neighboring communities with an aerosol spray of *Serratia Marcescens,* a type of bacteria the military believed was harmless. If it was, why the experiment? If toxic, then such an act was grossly callous. In 1977, the Army stated to a Senate health committee that it had conducted 239 tests between 1947 and 1969, of which eighty included disease-producing organisms. The areas covered were San Francisco; Key West and Panama City, Florida; New York City; and Washington, D.C. The only known fatality was in San Francisco; other persons developed *Serratia* pneumonia but survived. But since there had been no public disclosure of those tests, physicians and pathologists were not alerted to the dangers.

Undoubtedly, the most studied volcanic eruption—with the exception of those of Krakatoa and Pompeii, which were done in retrospect—has been the earthquake-volcanic explosions of Mount St. Helens in Washington in May 1980. Clouds of ash spewed 63,000 feet into the stratosphere. Although it may have produced magnificent sunsets hundreds of miles away, the airborne ash reached the Atlantic Coast within a week, polluting the atmosphere. Millions of tons of hot mud flowed down Toutle River, destroying homes and bridges in its path. Within a short time a massive dam a mile long and twenty stories high backed the water near Spirit Lake fifty to 100 feet in depth. It is feared that when this mud-ash dam breaks, the towns below may be eliminated.

The ash has fallen in adjoining states; and in certain areas of Washington, approximately fifteen to twenty tons per acre have layered the soil. In decades to come, it may enrich the earth; but this is small comfort to farmers whose crops have been lost, or to Weyerhauser and other timber-holding companies that have lost over a hundred thousand acres of magnificent trees. In 1981 Dr. Peter Baxter of the National Center for Disease Control stated that the sandlike particles in volcanic ash irritate the lungs and increase the possibility of silicosis. Asthma, bronchitis, and susceptibility to lung disease have also occurred in the area.

Few persons can expect to live or work in an environment that is totally safe. The important factor is that the individual be informed of the risks and that the risks be kept as minimal as possible. Certain atmospheres will contain trace metals, others pesticides, solvents or allergens, and still other atmospheres may contain hydrocarbons, carbon monoxide, or the oxides of nitrogen. The strip miner will be exposed to clouds of dust that may contain dangerous particulates. It is important that the worker be educated as to dangers and protected as well as possible. The author has worked with *Clostridium botulinium*, which produces the most virulent poison known to man. He has also worked with some of the most dangerous strains of *Staphylococcus*, *Streptococcus*, and other disease-producing organisms—any of which, if inhaled or carried to the mouth, could produce disease or death. Since he was aware of the dangers, however, he was able to use safeguards and attain retirement years.

We should remember one thing—as many persons are destroyed each year in the United States by pollution as were killed in the entire Korean War. Let us be certain that we all try to reduce the millions of tons of dangerous gases and particulates that fall upon the United States each year, and not make this nation a befouled prison, like that of which Oscar Wilde wrote:

> The vilest deeds like poison weeds
> Bloom well in prison air.
> It is only what is good in man
> That wastes and withers there;
> Pale anguish keeps the heavy gate
> And the Warder is Despair.

The Foods We Eat

A bowl of porridge piping hot
With salt the Highland's weakness
The Sassenachs have liked it not
But this is due to meekness.

—Milo Don Appleman,
Songs of the Highlands

The foods we eat depend to a great extent on ethnic habits, origin, customs, availability, expense, taboos, and religion. Most of these are so commingled that it is difficult to grasp a single thread in the skein and unravel it to roll it into a smooth ball. One reason is that with the tremendous amount of international travel, persons are exposed to menus of different cultures; once they have become accustomed to the new foods, they are prone to adopt them with no difficulty. Many of us have eaten rattlesnake which, when broiled correctly, is as delicious as chicken. Escargots (snails) in garlic butter stuffed in the shell are a luxury, as are frogs' legs and abalone or other shellfish. To certain groups, however, these foods are taboo, usually based upon religious teachings.

The foods we eat are also based upon our climate, new surroundings to which we are unaccustomed, and the introduction of new methods of treatment of foods that were unknown previously. Interestingly, milk and butter have been disliked by many Chinese and Malays, among whom soybean milk, particularly when made into curd or cheese, is accepted.* This, in part, was due to the lack of, or scarcity of, the milk cow; although in some areas the water buffalo was a beast of burden and valued highly, but not as a milk-producing animal. The Chinese immigrants to the United States brought most of their native foods, the popularity of which is shown by the number of Chinese markets and restaurants in the major cities. Although the great-grand-children of many of these immigrants are now eating "junk foods" in the high schools and universities, most of them seem to prefer their native foods when available. The same is true of the Japanese, although there have been fewer areas in which this could be studied en masse.

Italian immigrants changed their food habits slowly, and, because of this, introduced many foods that have become major market items and main features in restaurants of ethnic origin. Most Eastern Europeans immigrating to this country tended to adapt to the foods present, although their heritage remains, in the many forms of Polish sausage and Hungarian goulash. Since the United States was settled to a great extent by English, Scots, Swedes, and Germans, quite early a variety of foods was established. Probably the one item that never became established was the preference to have one's meat, game, and birds "high," which is defined as one step removed from being putrid.

When the Pilgrims first landed, they had an insufficient supply of food and quickly adapted to corn (Zea maize) of this country instead of their corn (wheat) of Europe. Pumpkins and root crops were also grown and with these, those who survived the first winter learned how to live. Many made pemmican following the example of the Indian, pounding dried meat, melted fat, and dried berries together to produce a food that was stable for long periods of time and high in proteins, fats, and vitamins.

Blubber was eaten by the Eskimo and this fat replaced the starch that might otherwise be needed, and some of the vitamins. The Masai tribe are famous as being herders of cattle, of which they use the blood, rather than the meat, as a part of their diet.

Fish is disliked by some of the African tribes. Other coastal tribes have become fishermen, and marine foods form a large portion of their diet. Chickens and eggs are not liked in certain portions of Africa

*Many ethnic groups, because of an enzyme lack, cannot handle the lactose in milk.

and India, whereas in other portions they are important. In the United States, clam chowder started as a food in the Northeast, but is now universally eaten. Chili had its origin in Mexico and the southwestern United States, but it has become a commonplace food. Black-eyed peas and grits have moved from the Deep South to almost all markets.

Instinct is not a good judgment on which diets can be based, for the untrained individual turned loose in the woods would be as likely to gather a mixture of poisonous and nonpoisonous mushrooms as of nontoxic varieties alone. This will be discussed elsewhere.* A tremendous difference occurred in the diet of the Central and Deep South as contrasted with that of Mexico. Persons living in the hill country of Kentucky—who depended for their diet upon corn or corn flour products, fatback (or salt pork), and the greens they gathered in spring and summer—frequently died of pellegra, a vitamin deficiency disease known as black tongue. This was unknown among the Mexicans, where the corn was soaked in dilute lye, rinsed, dried, and then ground into masa flour. The lye soaking liberated niacin, which was otherwise firmly bound in the corn and not released to the human using it as food. Instinct did not help the Kentuckians.

Greenland was settled around A.D. 985, but within 500 years the colony had died out. These people subsisted largely on fish, and yet the examination of the skeletons indicates that most had rickets and died as the result. If instinct were of value, the people would have eaten the livers of the fish and there would have been no rickets.

From the standpoint of religion, Buddhism has few taboos. Theoretically, most Buddhists are vegetarians but many are not, and this is not a cardinal point of the religion. Christians have no definite food habits, although for some centuries they carried over a number of the taboos from the early days of Christianity. One of the last of the taboos to be eliminated was that against the eating of horses. The pagan and vandal horsemen against whom the Christians had fought ate horsemeat, and during the eighth century, Pope Gregory III confirmed the ban against eating horsemeat to make the distinction between the Christian and non-Christian more clear.

There is neither sufficient time nor space to discuss the complete early Mosaic sanitary code, which included food habits, but some of these portions were of great interest. An edible animal must have both a

*See Chapter 5, "Foods That Can Kill Or Injure Us."

cloven hoof and chew the cud. Thus the ox, sheep, and goat were usable as food, whereas the camel, which chews a cud, the horse, swine, dogs, and many other animals were forbidden. Fish must have both scales and fins; thus the delight of eating jellied eels and many other seafoods was never known. Jews could not eat any animal dead by means other than slaughter, but they could give it away or sell it to foreigners. The orthodox Jew could not use milk and milk products and meat at the same meal, nor use the same dishes for the other foods. Consequently, it became necessary to have two sets of dishes in the kosher household.

The forbidding of the use of pork was a sensible taboo, as many swine probably had trichina. A number of writers at present do not, however, believe this was the basis of the regulation.

Orthodox Jews eat unleavened bread during the week of the Passover. This is a memorial of their flight from Egypt, and at one time those not observing this ceremony were put to death. The High Priest of Jupiter in Rome ate unleavened bread. The reason is somewhat more obscure, but it probably was a "sin" offering so that the god should have only unspoiled food.

The Moslems adopted a number of the taboos of the Jews. Pork was forbidden, although camel flesh could be eaten. Alcohol was, and is, taboo, and during the month-long feast of Ramadan, nothing—not even water—passes the lips from sunup to sundown, at which latter time feasting begins.

Once a food has become taboo, it is usually very cheap as there are few takers, for most persons would not wish their friends to consider them so poor they must eat a forbidden food. Even though a taboo may be imaginary, it holds as strong a grip on the population as if a physical reason existed. In many areas lungs are not eaten, but kidney, liver, and sometimes spleen are. Even the intestine serves as a casing for sausages. The taboo against lungs probably grew from the observation that these organs were particularly associated with disease, and today we know that they can be massively infected with lungworm and the organisms causing tuberculosis.

The statement is often made that millions of Americans suffer from undernourishment. This is not quite factual, for within this country even persons on relief and on the food stamp program have sufficient funds to give them enough calories to prevent undernourishment. What does happen is that we do have some millions suffering from malnutrition, which is the lack of sufficient proteins or vitamins to maintain an adequate diet. I have seen children neglect or refuse to eat a balanced school lunch and instead satisfy themselves with a candy bar, a coke, and a small sack of potato chips—a diet which is all carbohydrate with some fat, and woefully lacking in protein and vitamins. I have also stood in line many times in large markets where the food stamps were used to pay for a six-pack of beer, a carton of cigarettes, and a lot of junk food.

This country contains approximately two to three million vegetarians who can secure an adequate diet by carefully balancing the carbohydrate, fat, and protein in the foodstuffs they purchase. It is much easier, of course, if the individual is not a strict vegetarian but incorporates eggs and milk in the diet. These latter foods, however, contain cholesterol and fats with high melting points; for this reason they, like mutton and beef, are eliminated from the diet of individuals avoiding high fats, such as in the Pritikin diet.

We also have a number of food faddists who wish only to eat "natural foods." This is a difficult area to define. The most knowledgeable of the group mean that they wish their foods free of pesticides and herbicides, which is an intelligent desire. Some persons mean that no food additive should be present in the product in any degree. This is, of course, impossible; since sugar, salt, pepper, mustard, baking powder, and many other condiments such individuals would add to a foodstuff if they were making it themselves are present in the market foods. Even brown sugar is usually not a natural foodstuff, but a white sugar flavored with caramel. Raw sugar can be obtained, but this has few advantages over white sugar since all sugars are *cariogenic*—that is, the microorganisms in plaques on the tooth can convert these to acid—and can cause cavities. Raw sugar does contain more proteins and extraneous matter, however.

There are few occasions in which the so-called natural foods demonstrate any advantage over other foods. I was vice-president of a large laboratory that examined many so-called natural foods, and it was found that most of these were the same as any other foods. A number of years ago, many of the sellers of natural foods purchased the products that appeared to be the largest and most perfect at the great food markets of Los Angeles. These had the same amount of pesticide residues as any of the other fruits or vegetables. There were no insect marks or signs of invasion. A few of the more intelligent customers realized that untreated fruits and vegetables would probably have been invaded by some of the natural pests. At once, the owners of some of the natural food stores began to purchase cull fruits and vegetables from the great central markets and charge high prices for these as untreated or natural foods. This does not mean that all of the health food stores were involved in a racket, but some were. Others tried conscientiously to purchase fruits or vegetable products that were pesticide- and herbicide-free. Unfortunately, drift from adjoining sprayed fields on some occasions contaminate even such produce.

At the present time the Food and Drug Administration (FDA) attempts to do a good job insofar as labeling of products is concerned. The consumer should read the list of ingredients of any product purchased, as these must be listed in order of the largest amount to the smallest. Usually, rye bread will be listed as containing wheat flour, rye flour, etc. This informs the consumer at once that wheat is the primary flour concerned.

Unless the homemaker has sufficient time to purchase every ingredient to prepare each part of a meal, she will use convenience foods to some extent. Whether these are dried, refrigerated, or frozen, there will be certain additives present to preserve the smoothness and freshness of the product. Actually, they can be of benefit to the consumer, for without certain additives the products could become rancid or lose flavor rapidly. For generations, butter has been colored with annatto, a yellow plant pigment, to give the same yellow color as is present in butter made in June when the cows are fed lush grass. Many persons accept this, but object to using the same color in margarine.

All foods that contain fats to any appreciable extent usually have BHA (butylated hydroxyanisole) or another antioxidant present. Many food products contain phosphates or other chemicals that add to the smoothness or keeping of the product. As long as these compounds are not toxic or carcinogenic, there is no objection to their usage. Just recently, however, a pie maker was sued for producing cherry pies made mostly of grapes. This was not harmful, but it was fraudulent; it was stopped and the manufacturer fined. The FDA demands that products containing insects or insect fragments be destroyed. There is a slight tolerance in certain spices and foods that are easily invaded. Recently one of the largest chain stores burned 650,000 Easter baskets from Taiwan because they contained spider eggs.

The division into natural foods means little when we read the list of ingredients and discover that a 100 percent natural cereal may contain 20 percent sugar; products such as yogurt and fruit juices, which state they are naturally flavored, list artificial flavors among the list of ingredients. Natural-style potato chips are made of potatoes sliced with the peel remaining on the potato, but are usually fried in oils high in antioxidants. Thus, there is little difference between these and any other junk food. Beers that are listed as "natural" have been treated with several chemicals to doctor the product. Whether this is good or bad makes little difference. The main point is that the term "natural" means almost nothing in a processed food product. Few products fit the author's definition of "natural," since even yogurt is made from skim milk with dry skim milk solids, flavors, and other materials added. Even a powder such as Tang, which one might presuppose to be made from powdered orange juice, contains no fruit juice but has many other ingredients flavored with some orange oil.

Seafood is not necessarily a simple ocean or lake product, since we can raise lobsters under artificial conditions, crayfish in rice fields, and even use hormones to modify nature. Thus, tilapia, a fresh-water fish, can be treated so that the young become males, which grow much faster and larger than the females.

In India today, most Hindus do not believe in the slaughter of animals and, as a result, the sacred cows can wander in the fields and destroy an appreciable amount of grain crops. Two states, West Bengal and Kerata, have a sizable non-Hindu population and slaughter those animals. In other parts of India the cow may be used as a source of milk and of fertilizer, in addition to being a beast of burden. In 1979, Vinoba Bhave started a hunger fast in an attempt to force the two recalcitrant states to form Goshalas, which are old-age homes for cattle, of which more than 1000 exist in India at present. The typical Hindu does not kill rodents; as a result, large portions of his harvested or purchased crops are destroyed.

In Egypt in 1981, tremendous numbers of rats were destroying food crops. These "super-rats" stripped grain, tomato, and vegetable fields, invaded warehouses, and climbed orange trees to devour their fruit. Their natural predators had been destroyed in the previous years of war in the Mideast. Approximately six million acres have been invaded.

Many countries of the world worry about the fact that many of their foodstuffs are imported. Japan, a progressive country in 1981, imported 60 percent of the foodstuffs for its 116 million people. It imports 94 percent of its wheat foods, 91 percent of its soybeans, and almost 100 percent of its feed grains. Japan is one of the best-fed countries of the world, with a per capita consumption of 2500 calories a day. An interesting point is that the Japanese are changing their diet, substituting sandwiches and noodles for the traditional rice.

There are many conflicts relative to foods in progress throughout the world at the present time. Seaweed, such as kelp, has been touted as a source of protein by certain groups. Although these plants are rich in minerals and vitamins, the usable protein level is low.

Some groups wish to produce finished foods with high acceptability and low calories in the United States for dietary purposes. We see examples of these on the market, in which a single can of flavored skim milk contains 200 to 250 calories. Almost all persons are in agreement that mother's milk is better for babies than are formulated milks; outside of the United States, formulated milks are usually poor sources of food for babies as they are overdiluted to save money and are also retained under unsanitary conditions.

The Food and Drug Administration—which has performed a good service in requiring the listing of ingredients as to their proportion in food, and statements of the proportion of RDA (Recommended Daily Allowance) of vitamins, minerals, protein, etc.—finds itself in a rather ridiculous position in the use of fish flour. This product is produced by extracting the protein of cheap, whole fish, such as menhadden and hake. Other fish also may be used. The department discourages its use to enrich the protein content of foods because the whole fish, including intestines and so forth, is present. Yet many consumers, without protest, will eat raw oysters or other shellfish which contain the entire gut. In the fish flour, the solvents and temperature have inactivated any bacteria, which is not true in the case of raw oysters.

About 90 percent of all of the food consumed in the United States is produced here. The exceptions include many items such as coffee, tea, cocoa, certain fruits, fish and shellfish. Using this country as an example, as income increases, the total expenditure for food increases. Sir Samuel Garth said of those of modest means:

> Hard was their lodging, homely was their food;
> For all their luxury was doing good.

In the higher-income groups more is spent on soups, vegetables, potatoes, sweet potatoes, bakery products, meat, cream, and cheese than in the low-income groups. Less flour, fats, oils, eggs, sugar and sweets are purchased by the former.

The per capita American diet increased from 1430 pounds of total foodstuffs in 1960 to 1470 pounds in 1981. In 1979 pounds per capita were approximately as follows: beef 80, pork 65, poultry 62, fish 18, cheese 22.5, fats including butter 61, processed vegetables 65, processed fruits 58, cereals and flour 150, and sugar 138. Soft drinks averaged 36 gallons per capita. There was a decrease in eggs, milk, fresh fruit and vegetables, and coffee during this same period.

When newer methods of preserving foods first came in, such as canning, refrigeration, and freezing, these were used by the income-wealthy groups who now purchase farm-fresh produce rather than frozen.

Antioxidants such as BHA have been mentioned as a means of preventing rancidity in fats. Many others exist and are necessary if a food is to be kept on the shelf for any period of time. Certain harmless preservatives, such as the propionates and sorbates, are necessary to prevent the molding of bakery products and similar materials. The word "harmless" is used deliberately, for in the production of Swiss cheese, propionic acid is produced naturally by the organisms necessary for the manufacture of this cheese, and the amount in the breadstuffs lies within this percentage, which is 0.2 to 0.3 percent. There are certain preservatives, flavors, and colors that should be permitted in foods to lengthen shelf life and to render the food attractive. Other colors and preservatives should be barred. The FDA is doing a good job of sorting this out.

It is obvious that were it not for the convenience foods on the market, the homemaker would have to spend at least eight hours each day in making purchases from the different markets, as is true in Iran and other countries, and compounding these into two or three meals that would satisfy her family.

Certain countries of the world are taking advantage of breeding new grains for increased yields. Mexico and Canada have produced a new variety of wheat that almost tripled the yield in Mexico. Taiwan has worked with other countries to double the yields of rice. Daniel Benor of Israel has introduced new systems of cultivation into India that have increased the production of certain crops. But much of Africa and of the Orient still use more primitive methods and poorer seed grains.

It is interesting to travel through time almost two hundred years ago when England and France were locked in a life-and-death struggle to determine whether Napoleon would rule all of Europe. The French built better and faster ships than the English, but were constantly beaten. The diet of the French sailors was almost unchanging, consisting primarily of salt beef, salt pork, and weevily biscuits. But the previous British admirals—many of whom had been little better than pirates—had discovered an important fact. If the juice of a lime was squeezed into the sailor's ration of grog (rum) each night, he was a healthy, fighting brute. Without this, scurvy, a disease caused by lack of Vitamin C, developed. The sailors became ill and eventually died unless they had fresh fruit and vegetables. As a result, in battle after battle the English navy proved superior, and was finally able to blockade France.

When this occurred, France was cut off from its supply of sugar from the Indies. Napoleon believed that sugar could be produced from beets. Not only was this agricultural change made, but high sugar-containing strains were selected that would yield large amounts of sugar.

Napoleon soon realized that his fleet could not match that of the English insofar as diet was concerned, and offered a prize to anyone who could preserve vegetables, meat, and fruit for his navy. This prize was awarded to Nicolas Appert in 1810, a confectioner who developed the process of canning. He sealed his products in jars and immersed these in boiling water to destroy the microorganisms. Experiments showed that the water had to be a brine of calcium chloride or sodium chloride to raise the temperature of the boiling water considerably higher than boiling temperature. The science of canning, which now uses pressure steam, developed from this early work.

In Britain during World War II, there was a severe shortage of sugar for candy and as a result, the children had much better teeth than either before or after the war. In order to furnish Vitamin C, concentrated orange juice was purchased and the vitamin content reinforced with that extracted from rose hips from Scotland. Wheat flour was in short supply, so 10 percent corn flour was added to this. To make the bread appear whiter, chalk was added, which again was good from the standpoint of tooth and bone structure.

Although the world grain production has increased markedly in the last twenty years, the amount of food available per capita has decreased due to a proportionately larger increase in population. This is not a constant factor in each country since the greatest percentage decrease in food available per capita has been in Latin America, Africa, and the Orient. There has been an increase in food available per capita in the Near East, North America, Western Europe, and probably Eastern Europe. In the underdeveloped countries food production has remained quite constant at a low level.

In the 1880's twenty man-hours were required to produce and harvest an acre of wheat. By 1944, the time required was reduced to a little over half of this time; and at present, due to mechanization and uniformity of crops, less than five hours are required. The great reduction of time utilized per acre of ground results from the introduction of new varieties that not only give greater yields per acre, but also grow to uniform height so they can be harvested mechanically with ease. The introduction into the United States of hybrid corn, in which there were two ears per stalk rather than one, meant that with adequate fertilization, yields on farms which at one time produced forty bushels per acre now produced eighty to 120 bushels. Soybean yields and those of other crops increased even though the production time-per-acre decreased.

It is necessary to use mechanization on the present-day farm, although one should keep in mind the amount of energy needed to produce that machine and deliver it to the store, and eventually to the farmer. To compare energies, it is necessary to do this in terms of an acre or a thousand acres. On small farms which have practically gone out of existence, there was an advantage in the use of horses and mules since part of the energy they consumed was returned to the land in the form of manure. A tractor gives only exhaust gases. However, because the small farm has given way to agri-industry and large tracts, it is necessary to use the tractor, the combine, the corn picker, and other automated machinery.

The definition of food should be reemphasized. Undernourishment is the lack of sufficient calories to support the needs of the body. This is a quantitative factor and varies with the amount of work performed. Malnutrition is a qualitative factor and usually refers to a lack of protein, although it may also refer to a lack or deficiency of vitamins. When protein is greatly deficient, the individual will develop kwashiorkor and die of this as readily as from undernourishment.

Kwashiorkor is the most important nutritional disease of children in the age groups from several months of age to four years. The most common clinical signs are failure to grow, edema, atrophy of the muscles, misery, and loss of activity. There is enlargement of the liver and anemia. Usually poor appetite and diarrhea are present; this can

go so far as to lead to almost complete starvation of the skin and
bones. The disease is present in children weaned to a diet of almost
complete maize in Africa, Central America, or South Africa, or to a diet
of cassava, plantains, and sweet potatoes in Central Africa. If the child
is placed on a good diet containing milk and abundant protein, he or
she will recover relatively rapidly.

This disease is not associated in any way with the Pritikin or other
diets frequently used by heart patients or those individuals using these
diets in a preventive manner. These diets depend upon the use of
complex carbohydrates in vegetables, including garbanzo beans, with a
restricted amount of simple sugars. The vegetables used in the entire
diet are augmented by protein from legumes or from animal foods with
low melting-point fats, such as fish and chicken.

Deficiencies of various vitamins cause a number of diseases, so
many that the individual interested should study a book of nutrition.
An individual attempting to supply his calories through a large intake
of alcohol can be both undernourished and suffering from malnutrition.

According to most nutritional tables, an individual should have
an intake of 2800 calories, including 300 calories of protein, per day.
This recommendation is too high for the average sedentary American
and too low for those engaged in hard muscular activity. North
American and Western European populations receive more than the
suggested amounts of calories and protein, and those of Latin America
and Japan approximately the correct amount, although in the latter
country there is less meat and more vegetable protein. Eugene Pottier
wrote:

> Arise, ye prisoners of starvation,
>   Arise, ye wretched of the earth,
> For justice thunders condemnation—
>   A better world's in birth.

The preceding applies to the many countries of central Africa, and
to India and Pakistan, where the intake is approximately 2000 calories
and the protein intake is 160 calories. It is to be hoped that these areas
will produce more foodstuffs in the future, or manufacture goods that
can be exchanged, to improve the welfare of their populations.

It is dubious if the countries of the world that can furnish their
populations good standards of living, insofar as food is concerned,
should assist a country that contains numerous starving persons and

where the population rate of that country is increasing rapidly. To place numbers to this; is it less moral not to furnish food to India in 1984 and permit twenty million persons to die of starvation, or to allow 200 million persons to die in A.D. 2000 when we could be of little assistance to them—or to Bangladesh or any of the Third World countries—due to our own increased population and diminishing food supply?

Even the Soviet diet is worsening, for in 1982 they expect their fourth straight poor grain crop and will not be able to furnish great assistance to their allies except through grain purchased from the United States, Canada, and Australia.

The protein and calories capable of being produced depend on the use of the land. One acre of good farm land will produce approximately forty-five to fifty pounds of beef protein per year, seventy-five to eighty pounds of milk protein, 450 pounds of soybean protein, and 650 pounds of alfalfa protein. In terms of calories, 4.5 million can be obtained from wheat grown upon an acre of good land, eight million from potatoes, and thirteen million from sugar beets. It is obvious that in the nations of the West both protein and calories are "wasted" on a meat and bread diet. Undoubtedly, a time will come when it will be impossible to produce meat animals on a limited amount of land, and most of mankind will subsist upon cereals, legumes, and root crops. Only small amounts of fish will be available for protein because of our present careless neglect of our fishing.

In agriculture, it is necessary to use a number of different food plants and to rotate these so that the same crop does not grow upon the land more than one or two years in succession. In the United States, in certain areas rotations of corn, soybeans, and wheat are common—in which case a legume crop, which harbors bacteria that can fix nitrogen from the air, is sown with the wheat. In other countries, rotations of crops may consist of barley, potatoes, and clover. Any crops that are suitable to the soil and to the climatic conditions may be used. If a single crop is grown upon the soil year after year, usually micro-organisms, nematodes, or other organisms that can destroy that crop will develop. This use of a single crop is called monoculture and is to be avoided.

It should be pointed out that those individuals who are natural food addicts to the extent that they do not believe any commercial chemical fertilizers should be used may be unaware of the process involved in soil nourishment. Normal food crops that are not legumes only take up nitrogen in the form of nitrate. If one manures the soil, the protein of the manure is converted to ammonia by many groups of bacteria, and then through nitrite to nitrate by a limited group. Using compost or manure for the organic matter does help keep the soil in good condition or "tilth." I do wish to indicate, however, that chemical fertilizers can be purchased which contain the nitrate in readily utilizable form and, if the correct ones are obtained, these will also have the trace elements such as manganese, cobalt, ad infinitum, that are also present in manure.

In Ireland in the 1840's, potatoes were grown on most of the tillable land as they furnished more calories than any other crop adapted to the soil. As a result, after a number of years a population of pathogens developed which destroyed the potato crop. Since there was no secondary crop to any extent, many of the people starved or were forced to emigrate. Ireland lost approximately half of its population by death or emigration. Growing of a single crop also removes the nutrients needed by that particular plant to a much greater extent than would occur if crop rotation were used.

In certain lands of Hawaii, where pineapples are grown, a system of monoculture develops. This is not desirable, but the soil scientists usually fumigate the soil to destroy pathogens and frequently hold the toxic nutrients in the soil for some time by laying sheets of Kraft paper or roofing paper between the rows to hold the toxic gases in place.

A limited population can be supported upon the planet if meat, eggs, and milk were to form a major portion of the diet. If the cropping system is changed so that the legumes (such as soybeans), rice, corn, wheat, or other cereals were substituted, a much greater number could exist. Speculation is that possibly the earth could support ten billion persons on the first type of diet and possibly thirty billion upon the legume, cereal, and root crop culture. If man finds a way to grow algae in large masses wherein these usually single-celled plants could obtain their energy from sunlight, producing carbon compounds by the green pigment chlorophyll, the number of persons that could be supported might increase into the hundreds of billions if living space could be found for them. A difficulty would still remain in that experimental psychologists have shown that if a rat population is permitted to increase to an extreme limit, these rodents eventually become neurotic, antisocial, vicious—and may die or kill each other. It would probably be a much worse situation were humans crowded to such an extent.

Many persons have thought of the water sources, particularly of the oceans and the seas, as being almost a limitless supply of food. No greater mistake was ever made. Approximately 90 percent of the ocean is known as open ocean and is relatively infertile. It does not produce the plants upon which smaller marine life feed to support the larger fishes. As a result, less than ten percent of the fish catch landed comes from this area. The coastal zones and continental shelves are 9.9 percent of the ocean area and produce 37 million metric tons of fish per year. In certain areas of the ocean there are upswellings. These may be due to the strong trade winds that blow the surface waters out to sea, with the nutrient rich bottom waters rising to the surface. They may also be due to deflections of deep currents of cold water rising to the surface when an obstacle is encountered. These conditions are so ideal, insofar as nutrients are concerned, for although they constitute only 0.1 percent of the ocean area, they produce over 23 million metric tons of fish annually.

In areas such as Chile and Peru, the nutrient-rich waters rising to the surface produce phenomenally large crops of anchovies upon which mackerel, tuna, and seabirds feed. Some of the latter were responsible for the rich deposits of guano that for years were used as a source both of fertilizer and gunpowder.

In the Antarctic waters, the rich nutrients produce large crops of shrimp-like crustaceans known as krill, upon which certain of the whales feed. One blue whale was found to have engulfed two tons of krill, which is the current record.

The reason man cannot depend upon the oceans for an unlimited supply of food is that already certain areas are being overfished and there is a shortage of certain species. In California, the sardine industry was destroyed by overfishing and by kelp cutting. The anchovy that replaced the sardine is already decreasing in yield.

Fish farming, other than crawfish farming in rice fields, will not yield the same food and price yield in the United States that can be obtained from appropriate crops in good agricultural soils. Fish farming, both for food and for sport, has been undertaken in certain areas. This could well be done in any large pit left by coal mines that were abandoned in West Virginia, Kentucky, Illinois, and other states. Such states have sufficient water to keep the pits filled and are in areas where even in hot summers the bottom waters are cold. These lakes need fertilization with a nitrogen and phosphate fertilizer to start the growth of phytoplankton upon which the small zooplankton will feed. Approximately two weeks after fertilization, the lakes can be stocked with native fish, particularly bluegill, crappie, and bass. A finely chopped fish food or horsemeat must be fed the first year and crops are greater if feeding is done each year. In the Philippines and other parts of the Pacific area, fish farming is an important industry.

Edible fish that are not eaten by man directly are ground and processed into fish meal to feed lower animals. As a result, only a portion of the protein and calorie supply is recovered when these animals are used as food, since part is lost in respiratory and metabolic activity.

Many of the major countries of the world, including the United States, Japan, Russia, and East Germany, have factory ships that are accompanied by trawlers. The factory ships locate in an area of good fishing and the trawlers bring in their catch each day or two. As John Heywood said, "All is fish that cometh into the net."

Upon the large factory ships there are facilities for filleting and freezing the best fish. Other fish are frozen whole, but all remains and coarse fish are processed into fish meal. All fleets are not accompanied by factory ships, but the depredation is such that, due to all combined, the ocean is overfished and the point will be reached where replacement is not practical. In addition, certain ships use nets that are so fine that fish that should be able to escape cannot do so. Many porpoises are killed in the trawling operations.

I was in Scotland during part of the years that Britain and Iceland had difficulty about foreign vessels fishing for cod off the coast of Iceland. Eventually, conditions caused the "cod wars" to erupt, but a peaceful settlement was finally made. In the North Atlantic, however, the coastal waters have been overfished for cod, haddock, plaice, and even such coarse fish as hake.

At the present time, somewhat over 60 million metric tons of fish comprise the harvest from the ocean. Of this, approximately thirty-five million tons are used directly in human consumption, and twenty-five million tons are processed. Part of the processing is for fish meal to be fed to poultry and livestock.

The total annual resources of the oceans as a source of fish for man will amount to eighty-five to 100 million metric tons if all nations agree not to overfish and to use nets that will permit smaller fish to escape. In addition, the whole supportive community of life upon which fish feed is being devastated by the fine-mesh trawls. It is of interest to all that some of the nations of Latin America and of Africa have greatly expanded their catch and are beginning to rival that of the traditional fishing nations. The statement of Jonathan Swift certainly applies to the larger fish that prey upon the smaller until the smallest forms are reached:

> So naturalists observe a flea
> Hath smaller fleas that on him prey;
> And these have smaller still to bite 'em;
> And so proceed ad infinitum.

There is hope that man may develop foodstuffs from unusual sources. It is known that certain bacteria can be grown on hydrocarbons or using hydrocarbon gases as a source of energy. Although this is not an economical process as yet, it would yield what is known as single-cell protein when this is isolated from suspension of cells. Algae have been grown and used as a source of foodstuffs, using sunlight as the source of energy but otherwise in a mineral medium such as the bacteria.

Experiments have been made to produce protein and carbohydrates from leaves, agricultural wastes, and sewage wastes. It is possible that one of these methods might be sufficiently successful to relieve a portion of the drain on the world supply of calories and protein.

Beginning in 1974, the United States Department of Agriculture (USDA) and National Aeronautical and Space Administration (NASA) devised satellites called Large Area Crop Inventory Equipment (LACIE). These satellites could measure the land area producing almost any kind of crop. With these AGRISTARS® and others that measure the oceans, we are able to monitor the renewable and nonrenewable resources with a great degree of accuracy.

We may conclude from this discussion that—assuming all factors favorable to food supply so far as nutritional qualities and lack of tox-icity are concerned (and this is not always a safe assumption, as we shall shortly see)—mankind is confronted with several problems which must be solved if the epitaph is to be put off in time. The first relates to judgment in selecting foods necessary for health. The second relates to the intelligent use of resources to produce the maximum amount of food for an expanding population. Finally, we cannot destroy the areas upon which we depend for food, whether it be farmland (either from acreage reduction or improper usage) or our waters; if we do not preserve them, they will not preserve us.

Foods That Can Kill or Injure Us

This subject breaks down into a half dozen topics, each of which easily could fill a volume. However, such intensity of treatment is not required, since our purpose is to discuss together threats to our continued survival. We eat, if we are fortunate, several meals a day. As a rule, that which we consume is not life-threatening, at least if we exercise reasonable judgment. Hazards do exist, however, which must be intelligently evaluated, both by us as individuals and by governmental bodies exercising the police powers of public health and safety.

The first two categories, it will be seen, are based upon the physiology of the individual; the next four upon the substances in question:

1. Metabolic deficiencies. There are in excess of a hundred foodstuffs that are clean and wholesome, but which a given individual—perhaps because of an ethnic metabolic deviation—may be unable to utilize safely. Galactosemia is a condition in which a newborn infant is devoid of a certain enzyme* and, because of this, could suffer death, brain retardation, or other physical adversities if placed on a diet of cow's milk.

Approximately 70 percent of Jewish and Negro adults lack the enzyme lactase, and can handle beer better than milk. My own research, although not final in this area, indicates some deficiency on the part of Mexican and South American Indians. I have recommended substitutes for cow's milk in the form of soybean or sesame seed milk for such persons.

Tay-Sachs, and other disorders, result from genetic deficiencies.

2. *Allergies and sensitivities.* Allergies and sensitivities may exist in many nonfood areas, such as a reaction to medication, pollen, dust, wool, feathers—even to one's spouse. In the food area, normally such idiosyncracies (in the scientific sense) show up over a period of time. When repeated exposure to a given food or beverage produces an identical result, the warning flags should begin waving briskly, and parents should take heed.

Such sensitivity frequently exists with reference to any shellfish. Other persons react to eggs, milk, almost any protein-linked substance, or to particular fruits and vegetables.

Nutritionists are now seeing vastly increased numbers of hyper-kinetic (overactive problem) children whose conduct is traced to cereal or sugar sensitivity. A withdrawal of the foods in question has caused a vast change in such children's behavior.

One individual, an adult, was sensitive to corn in beverages; he developed migraine headaches if he drank bourbon, but none after he switched to Scotch.

3. *Toxic and dangerous foods.* We've previously been discussing healthful foods to which a small percentage of persons suffer a reaction. As Lucretius said, "What is food to one, is to others bitter poison."

But many foods may be toxic, or even fatal, to those with no idiosyncracies. Taking seafoods, for example—puffer fish poisoning occurs most frequently in Japan. It seems to be the result of the method of preparation of the fish and possibly the length of time the fish were not refrigerated. The symptoms develop rapidly, in most cases starting with tingling of the extremities, followed by numbness of limbs, lips, and tongue. Nausea, vomiting, respiratory distress, and loss of move-

---

*This enzyme is galactose-1-phosphate uridyltransferase. All first-rate hospitals im-mediately test for this and compensate by appropriate diet.

ment of the eyes is followed by extreme muscular paralysis. The death rate is approximately 60 percent.

The symptoms of moray eel poisoning and of Ciguatera fish poisoning are quite similar. In both, these include nausea and vomiting, metallic taste, abdominal pain, fever, paralysis, and other symptoms. The death rate varies from approximately 3 to 9 percent in each.

Many types of shellfish, such as mussels, clams, and oysters, can become poisonous from May 1 to October 1 because a toxic form of life called dinoflagellates serves as a major portion of their food. On the Pacific Coast, the species involved is *Gonyaulax catanella*; and during the dates specified above, a major portion of the California coastline is quarantined so that people cannot use the shellfish. Other species are involved on the southern and eastern coasts. The liver of a single mussel may contain enough toxin to kill from 1000 to 5000 mice.

There are many types of marine life that may be toxic in certain areas of the world and at certain times. Squid and octopus may produce illness in peoples of the Orient. The toxicity is known as cephalod poisoning and although severe, the death rate is a fraction of 1 percent. There is also a scombroid poison caused by the action of bacteria upon mackerel, skipjack, bonito, and tuna. Its action is histamine-like in nature and can cause headache, nausea, vomiting, hives, and other symptoms—but its death rate, also, is less than 1 percent.

One natural type of food poisoning—which has always been intriguing because of its possible relationship to the Biblical stories of Moses leading the Jews from Egypt—is known as "green quail" poisoning. The Biblical account states that when the people were hungry and wanted bread, Moses caused manna to appear for them. When they wanted meat, large flocks of quail descended before them. Because the people were greedy and unappreciative, the quail became poisonous in their mouths.

Quail migrate in the fall from Europe to Africa, and in the spring from Africa to Europe. At the time of the migration of Jews, North Africa was a fertile land of many trees including olive, hemlock, and others. Writers such as Pliny the Elder and Galen have mentioned that quail could be poisonous to man at certain times. The possibility exists that in the great migratory flights of quail from Africa, the first flights found the quail to be wholesome food. Later flights might have contained quail that fed upon hemlock seeds, which are nonpoisonous to the bird—but the meat of the birds has been proven to be toxic when fed to dogs. Hemlock can be a rapid-acting poison, and, as most will remember, was used to poison Socrates.

Certain natural foodstuffs contain substances known as pressor amines, which include serotonin, tyramine, histamine, and other compounds. These are found in very small quantities in a number of foods such as ripe tomatoes, passion fruit, bananas, plantain, and pineapple. Camembert cheese is high in its content of tyramine and should not be used by persons using parnate and a number of tranquilizers. Among certain African tribes that consume plantain as a major part of their diet, relatively large numbers of cases of endomyocardial fibrosis occur. It is ironic that this bulky food, which sweeps the intestinal tract clean so rapidly that cancer of the colon is practically unknown, has this side effect upon the heart.

Certain foods contain cholinesterase inhibitors. These are compounds that inhibit nerve impulses and occur in certain insecticides, such as the organic phosphate nerve gases and the carbamates. These substances would not be expected to be found naturally in foods. In potatoes that have grown above ground, and are green, however, the compound solanin, which is a cholinesterase inhibitor, is found. In West Africa the Calabar bean contains appreciable amounts of another type. These inhibitors can be extracted from cabbage, broccoli, peppers, pumpkin, squash, and other foods, but fortunately the amounts are so low they can be classified as inactive.

Another disorder can be caused by eating the uncooked broadbean or fava bean; in fact, attacks of favism have been produced by exposure to the pollen of the plant. And for heart patients familiar with coumadin, the anticoagulant, a like effect may arise from the related coumarin, once used in artificial vanilla flavorings. Sweet clover, on which cattle have fed, has caused hemmorhaging and death—this was endemic in the 1930's and 1940's in the Midwest. Persons eating the flesh of cattle feeding on sweet clover in which certain nitrogen compounds have become fixed could, inadvertently, be anticoagulating their blood.

Many of the plant foods we eat are goitrogenic and some of these properties can be transferred to man via milk of cattle fed upon these. The family *Brassicae*, which includes turnips and rutabagas, are goitrogenic, as are cabbage, broccoli, kohlrabi, and brussel sprouts. Fortunately, such properties, for the most part, are found in the seeds. Potassium iodide is added to table salt to supply any deficiency in iodine, and acts as an inhibitor to certain of these, but not all.

Other foods may be inherently carcinogenic. Coffee, or at least decaffinated coffee, is now suspect. As carcinogens are discovered, the Food and Drug Administration ordinarily releases information concerning them. Later, we will refer to one or two items most commonly encountered. For those natural food lovers who prefer old-time teas, it should be pointed out that sassafras tea should be a no-no—safrol, a bark derivative, is a carcinogen.

There are a number of foods which are ordinarily wholesome, but which may become dangerous because of molds or fungi to which they are susceptible. Since we had a peanut-growing, peanut-wholesaling president, it is appropriate first to mention aflatoxins, *Apergillus flavus*, to which peanuts are susceptible. While such toxins can readily be detected by laboratory tests—extraction and exposure of prepared glass plates to ultra violet light—the fact remains that such tests are not always run. Moldy peanuts are prone to develop aflatoxins, which are carcinogenic, and such peanuts may directly, or when converted to peanut butter, be dangerous. Careful food processors test the stocks they purchase; lawsuits can be costly.

In Great Britain, in 1960, a severe outbreak of disease broke out in turkeys. This so-called Turkey X disease was caused by feeding moldy peanut meal.

Much more dramatic, over the ages, has been the disorder called Saint Anthony's Fire, or ergotism in rye. Rye is a cereal upon which the fungus *Claviceps purpurea*, in unfavorable years, grows onto and into the grain itself; developing toxic and carcinogenic compounds, some twenty in number. Throughout the Middle Ages this was a common disease; although the peasants knew the grain was infected they had to eat it as it was their only food in most cases. Some of these compounds restrict the flow of blood to the extremities, the toes becoming black, like the fungus, and gangrenous. Usually, the rye of an entire geographical area would be subject to the disease at the same time. Beginning with the onset of A.D. 857, it became known as Saint Anthony's Fire, due to the blackened extremities of the victims.

The people would dance and scream in the streets, stating that they were burning. In some areas the entire population of a village would be wiped out. In mild invasions, symptoms would last until the rye was all used. The last major reported outbreak occurred in France in the early 1950's. Some reports indicated that the miller involved there did not even use contaminated rye, but simply used the sacks in which ergot-infested rye was contained; while this is partially true, it did not, in my judgment, exclude usage of some of the infected rye.

Other fungi produce carcinogens called mycotoxins. Their presence may be determined by appropriate tests. Even the mold *Penicillium* contains materials carcinogenic for the liver.

4. Dangerous additives. Whenever the term "additive" is used, it seems to be employed as an expletive. And no matter who is speaking, or writing, the Delaney Amendment of 1958 is cited, although rarely quoted. The Food and Drug Administration claims that its hands are tied by this law; this, industry denies.

Accordingly, before plunging into this subject, let us quote the amendment:

> No additive shall be deemed safe if it is found, after tests which are appropriate for the evaluation of safety of food additives, to induce cancer in man or animal.

While I hold no brief for lawyers, it seems that the FDA should engage reasonably clear-headed counsel to read these words slowly to its administrative heads, and think logically and cogently of the meaning of these words. Certainly the words "tests which are appropriate for the evaluation of safety" must contemplate usages reasonable in human consumption (or in pets, if pet foods are in question). It does not have in mind using dosages 4000 times that which a human would devour in his lifetime. There are few, if any, substances one might ingest—however wholesome they might be—which would not seriously affect human health if the effective portion were increased by several thousand times (and I am speaking qualitatively, not quantitively).

Such tests are not "appropriate" and the FDA should rethink its approach. Of course, if it still retains its current timidity, Congress should not be too proud to reword the amendment in a form reassuring to the FDA. While it is important to safeguard our citizens against noxious medications and foods, we should not be deprived of many "optionals" when all reasonable tests demonstrate safety.

As an example, let us take cyclamates. Cyclamates are sweeteners. They were used in foods and beverages which are ingested by mouth, not injected into the bladder. Reasonable usage should be permitted until "appropriate" tests demonstrate otherwise. Saccharin also is a blessing to many persons who cannot tolerate sugar; it is more reasonable to limit, rather than prohibit, its use.

The Food and Drug Administration should, perhaps, be more vigorous in requiring legible warnings as to the use of the various forms of sodium. With the millions of persons in this nation who suffer from high blood pressure, or teeter at the upper limits of normal, the warnings as to *salt* should be as attention-grabbing as those on cigarettes.

Food processors are lazy. They tend to put salt in everything as a taste palliative. Here, for example, is a package of Instant Quaker Oatmeal. It says: "(We've even put the salt in for you.)"*

Big deal! Who asked them to? That's a great convenience, isn't it, to save one the onerous task of raising, and shaking, a salt cellar. One would be too exhausted to face the day's labors.

If I sound sarcastic, that is intended. We are trying, by educational means and clinical instruction, to wean people away from excessive salt usage. Many use no salt at all, and their range of salt-free foods is often limited. Food processors should be required either to omit salt entirely, trusting the diner to add such ingredient if and as desired, or post a warning on the package in large red letters:

> WARNING: This package contains salt. The Surgeon General has determined that continued use of salt may be hazardous to your health.

The FDA has required, appropriately enough, that carcinogens be removed from liquid smoke flavors. So far as I know, however, it has never warned the public that eating a charcoal-broiled steak may be as hazardous as smoking two packs of cigarettes. And, make no mistake, smoked foods are carcinogenic. Finland, which uses the largest percentage of smoked foods, easily leads the world in deaths from stomach cancer; areas of Japan, where such consumption is also high, pay a heavy toll as well. On emigrating to the United States, and adapting to

---

*The information on this package has now been changed.

different foods, the gastric cancer rate in Japanese persons markedly diminishes.

The greatest current argument is whether or not nitrites should be barred from foods in which they are used as curing agents, such as bacon, ham, sausages, pastrami, and corned beef. Bacon fried at high temperatures can produce nitrosamines. Nitrosamines are powerful carcinogens for animals but at present can only be considered as possible, not probable, causes of cancer in man. Nitrites are beneficial in that they give the color, part of the taste, and part of the protection against *Clostridium botulinium*, which produces a virulent poison. At present, I believe the amount of nitrite in processing should be limited sharply but not eliminated. Interestingly, ascorbic acid seems to be an antagonist, so that individuals who have a glass of orange juice with their morning bacon may be relatively safe.

Much more study of the human diet should be done before nitrites are eliminated. It is well known that sausages using a salt-nitrite cure were used in pre-Roman times and up to the present. Since frying bacon may form carcinogenic substances, should all bacon be removed from the market? My answer, right or wrong, is "hell no." Let the FDA make the manufacturer imprint upon each package something like, "Frying bacon produces nitrosamines which are carcinogens. If you eat more than four slices daily it may be injurious to your health." Possibly more nitrite results from eating celery, radishes, lettuce, and certain other vegetables than is ingested from the eating of cured meats.

Another chemical over which there has been great argument is the use of diethylstilbestrol (DES). It had been found that if this was added to feed, cockerels could grow rapidly like capons and put on weight, since this is a female sex hormone. The chickens were to be taken off the treated feeds and put on normal feeds for at least seven or eight days before sale. Due to greed, and to get extra weight gain, some producers left the chickens on the hormone-containing diet longer periods of time. The carcasses contained DES. Thereafter, the use of such feeds was barred. The next approach was to implant a small pellet of DES into the neck of each chicken, just below the head. On slaughter, the head and neck areas were sold as feed to mink and fox farms. Some of these farms were almost wrecked, as there was still sufficient estrogen to prevent normal breeding. All use of DES was then barred for poultry.

The use of a DES pellet implanted into the ear of cattle was banned by the FDA on July 13, 1979. This might have been a mistake since it has increased the cost of beef, as cattle with the female sex hormone gain at least a half-pound more weight per day than others. Possibly what should have been done is to have a veterinarian remove the pellet and issue a certificate fifteen days prior to the time the cattle were sold. Suspicions were aroused when women whose mothers were given large amounts of DES to prevent miscarriage tend to develop cancer of the cervix and uterus. There is no DES remaining in the carcass meat, however, or even in the organs, if a full fifteen days have elapsed after removal of the implant.

Shakespeare said, "I am a great eater of beef, and I believe that does harm to my wit." DES can harm more than merely one's wit.

An interesting food additive is monosodium glutamate (MSG), which enhances the full flavor of meat, chicken, and fish. In very young animals, it can produce brain anomalies if large amounts are used. This material should not be used in baby foods since the only one to appreciate it would be the mother; baby doesn't care. Some persons eating at Chinese restaurants develop a condition known as Chinese Restaurant Syndrome due to the large amount of MSG in the foods. The symptoms are similar to histamine toxicity. MSG is used under various trade names. It is an irritant to the stomach lining, and the household would be well off to avoid such an enhancer entirely—either directly or as contained in various packaged foods.

Materials added by man to foodstuffs that are carcinogenic are relatively few in number. The Food and Drug Administration has what is called a GRAS list. It details substances "generally recognized as safe" and includes all of the common ingredients such as salt, sugar, spices, and a group of chemicals which can make food more palatable, or which have a definite useful purpose. Many of the dyes used years ago were removed from this list and their use in foods forbidden. All new compounds submitted for approval must have been subjected to rigorous testing, usually by a university or research institute on two or three different types of animals. If a substance that was used to keep fats from developing rancidity is tested and found to be safe, this will normally become a permitted substance.

Medications, of course, also are scrutinized by the FDA. It is now engaged in a campaign of requiring manufacturers to re-establish the utility of drugs long accepted as safe. When new medications are developed, the proofs required have become so onerous and expensive that all small research houses have been squeezed out of the market, which is left to the few moneyed giants. The FDA was badly frightened, however, by the terrible results following the use of thalidomides and MER29, and is determined that no such occurrences will again take place.

Since almost all medications have some adverse side effects (not in the category of thalidomides or MER29), it would seem better to permit the qualification of useful drugs somewhat more freely than at present. Adequate warning is given to prescribing physicians both by accompanying literature and through PDR (Physicians' Desk Reference), published annually with periodic supplements.

So far as hard drugs and tranquilizers are concerned, they should not be considered as "foods," although many Americans use them as freely as candy. The dangers of such usage are well known, as are those of excessive alcohol and tobacco, and will not be discussed in this book.

5. Tainted foods. Perhaps the most important source of danger in foods comes from microorganisms which can develop toxins, poisoning the foods. There are many types, but chiefly involved are *Staphylococcus, Salmonella,* and *Clostridium botulinium.*

It is common to say that cooking will kill all harmful bacteria. That is not necessarily true. In certain instances, even if the organisms are destroyed the toxins can remain potent. Other toxins can be destroyed by heat, but the organisms remain so potent that unless the food is refrigerated or frozen, new toxins can be produced.

Staphylococcus food poisoning is produced by bacteria which are not abnormally resistant. It is usually introduced into foodstuffs by individuals who have pustules or boils on the hands or face, or who carry the organism in the nasal pharyngeal area. The history of this organism is of interest. It was first suspected of being a food poisoning organism by Sternberg in 1884 in outbreaks involving cheese. The organisms were isolated in 1904 by a man named Denys and proved to be the cause of an outbreak in beef. Barber, in 1914, found that milk drawn from the udder of certain infected cows became toxic within a short time. In 1930, Dack continued much of the work on *Staphylococcus* food poisoning, as did Hobbs in England.

Approximately 15 to 20 percent of the normal population carry staphylococci on the hands and 50 to 60 percent in the nasal tract. When the organism enters an appropriate foodstuff, it can multiply sufficiently rapidly to produce the enterotoxin within a few hours at room temperature. Foods that are particularly attacked are cream-filled pastries; tuna, egg, chicken, or meat salads; and canned tongue or ham that is precooked but not retorted for sterilization, and is marked to be kept under refrigeration. In addition, the organisms can grow by the million per gram in gravies, soups, and cheese.

When I worked the "creamed chicken circuit" giving talks to organizations for the university at various clubs and organizations in towns, occasionally I would feel a gastric uneasiness within two to six hours after dining, at which time I would have violent nausea and vomiting. One does not need a stomach pump for *Staphyloccus* food poisoning, as nature has built in an excellent one. Fortunately, after about twelve to twenty-four hours, except for feeling rather weak and wrung out, I was well on the way to recovery.

My special bête noire was cream-filled pastries. Even at a dinner where a beef roast or baked chicken was the entrée, usually the desserts, particularly pies, were baked first, then removed from the ovens and allowed to stand in the warm kitchen for five or six hours while the remainder of the meal was completed. If the chef had pustules on his hands, and one or more of these broke open and were mixed with the cream filling, perfect conditions existed for the increase in numbers of these organisms and their forming the enterotoxin. This is not lethal, but sometimes for the first few hours one might wish it were.

Roasts are frequently handled by hand during slicing and placing the meat on trays. If this is used at once it is not toxic, but if one permits the roast and gravy to be unrefrigerated overnight this could be toxic also.

To prevent these enterotoxic conditions with staphylococci, large bakeries making cream-filled pastries either rebake these at once at a temperature where the organisms will be destroyed, or they refrigerate or freeze the products immediately. If the products are permitted to stand at ambient temperatures, so that the staphylococci grow to millions per gram, no amount of normal heating will destroy the toxins. I always insisted, when a food product was sent into our laboratory, that a special stain, called a Gram stain, be made and the slides examined under the microscope. Masses of round purple cells indicated that staphylococci had been present even though they did not grow upon laboratory media. This meant that the possibility of the enterotoxin must be entertained. If large numbers of thin, red, rod shapes were found, this would indicate that *Salmonella* food infection was possible. If large, purple, club-shaped rods were present, organisms that are known as *Clostridium perfringens* or *Clostridium botulinium* might exist in certain types of foods.

Foodstuffs that are macerated, such as chicken salad, ham salad, and cooked ground meat products, support the growth of staphylococci and other organisms more readily than do sliced meats or cheese, although these can be substrates for food poisoning organisms. In Los Angeles, regulations are quite strict for the large companies making sandwiches commercially. The workers have a large glass or plastic shield in front of them, which protects the foods should they cough or sneeze. Holes are cut in the proper position so that their washed hands and arms may pass through. If the worker leaves the line for any reason, he or she must wash his or her hands thoroughly before returning to place. Sandwiches are wrapped immediately and held as close to 32 degrees F as possible. The temperature range normally accepted is 40 degrees F to 45 degrees F. All foodstuffs must be discarded at the end of the day except acid foods such as pickles or pickle relish.

At one time it was difficult to trace staphylococcal sources of poisoning, as the presence of the organisms in the foods was not positive evidence. Kitten ingestion tests were not very reliable, nor were monkey feeding tests. Few human volunteers wanted to eat the suspected food. Now specific tests have been devised. In the most rapid method, what are called "fluorescent antibodies" are used. These are specific against each strain of food-poisoning staphylococci, of which several exist. Under a special type of microscope, bright fluorescence around the cells is seen. *Staphylococcus* food poisoning is one of the most common types but, fortunately, few fatalities result.

Botulism is a rare but highly fatal type of food poisoning caused by eating foods in which the toxin has been produced. A few cases of infant botulism, which is different, also occur. The organism was isolated from a spoiled ham by a man called Van Ermengem in 1895. The total number of deaths between 1900 and 1978 for all of the United States is less than the number of accidental deaths in any major city. The highest number of reported cases was 114 in 1977, whereas usually the case rate varies from ten to thirty per year. With efficient recognition and the rapid administration of antitoxin, the death rate is very low.

The organism *Clostridium botulinium* is found usually in the soil, but one type, E, may be present in water or in fish. The organism develops only in the absence of air and it is associated with outbreaks of mishandled foods. Usually these are home-canned nonacid foods such as beans, peas, meats, or fish, although some years ago a small, private label soup company was responsible for an endemic situation when its cream of mushroom soup was found to be the culprit. Foods in which the cans or jars have been processed by proper pressure cooking are quite safe. Occasionally, a leak in a can seal has permitted dirty water carrying the organism to enter, and a rare type of poisoning occurs. In Germany, botulism was associated with sausages, and in Japan and Russia with smoked fish.

Smoked fish from the Great Lakes produced an outbreak of botulism in Tennessee when those fish were held under nonrefrigerated conditions. The organism produces a spore or seed which is very resistant to boiling, but the toxin is destroyed by boiling for two or three minutes. This is exactly opposite to what occurs in staphylococci. The very resistant spore of botulism does not germinate or multiply under acid conditions. In addition, the organism is quite sensitive to nitrite, and this is one of the reasons I do not wish to see nitrite abolished from foods. I am afraid that cases of botulism in sausages—and in chopped canned meats such as Spam or Prem, which are safe now—might recur were all nitrite removed.

The various strains producing botulism grow readily at room temperature and one strain, E, the one associated most frequently with smoked fish poisoning, can grow in the upper ranges of refrigeration at about 37 degrees F to 40 degrees F, and at room temperature also. It is best to have these products marked commercially, "Keep under low temperature refrigeration or freeze."

The symptoms of botulism develop between ten to thirty hours after eating the poisoned product, although there have been cases in which three to five days passed before symptoms appeared. These are not evidenced by nausea and vomiting as in *Staphylococcus* food poisoning, but symptoms often include constipation and headache, followed closely by visual symptoms, primarily double vision. There may be drooping of the upper eyelid from paralysis and abnormal dilation of the pupil of the eye. Pharyngeal distress accompanies the above symptoms. The tongue feels coated and swollen and does not move readily. Speech is thickened due to developing paralysis of the pharyngeal muscles. The neck and lower back muscles may also be weakened.

The poison should be diagnosed clinically and antitoxin given. The responsible foodstuff may be determined by growing residual amounts of the food in appropriate media and injecting filtrates into immunized and nonimmunized mice. In the United States, types A and B are most frequent except where smoked fish is involved.

Infant botulism has been involved in more than forty-three cases in California since its first recognition in 1976. The sources of these cases have been vacuum cleaner dust, yard soil, and honey. In the infant there is extreme lethargy, feeble cry, and weakness. The infant is constipated. Infant botulism may result in a few of the cases of crib death called SIDS (Sudden Infant Death Syndrome). It has occurred in both breast-fed and formula-fed infants.

Clostridium perfringens, like *Cl. botulinium*, is an anaerobe although some strains can grow under reduced air pressure. This produces one of the most common types of food poisoning in the United States and the United Kingdom. This organism is at the midpoint or dividing line of food intoxications by preformed toxins and food infection. I believe that more than one factor is involved with this organism, but the fact that filtrates of the Clostridia grown in certain foodstuffs can rapidly produce symptoms in man indicates that an enterotoxin is present. The strains of organisms can also grow rapidly in the intestinal tract, which means that infection is also involved. The more of the food one eats, the more likely he is to become ill. As Shakespeare said; "They are as sick that surfeit with too much as they that starve with nothing."

Cl. perfringens is present in the intestinal tract of man and probably of all animals. When the animals are slaughtered the meat becomes contaminated. In addition, the human handler with fecally polluted hands, as well as flies, carry the organisms. Meats and gravies are readily contaminated, and if, after a meal, these are not promptly refrigerated, they can become the source of an outbreak. As with most foodstuffs, the rule should be to refrigerate below 45 degrees F, or heat above 145 degrees F, to prevent food infections. The symptoms are those of a severe gastroenteritis.

Bacillus cereus resembles the preceding organism insofar as toxicity is concerned. This organism is found primarily in the soil, however, and can be present in certain root crops. I have encountered only one massive case of *B. cereus* poisoning or infection and this resulted from steamed cull potatoes fed to cattle. In this outbreak, the organisms were present in millions per gram of potato and caused the illness or deaths of a number of animals.

Vibrio parahemolyticus is another borderline case between the food poisonings and food infections. The organism grows in the brackish portion of ocean waters, and the food poisoning is associated with the eating of raw shellfish and fish. Relatively few years ago, we believed this was confined to Japan, but, in the last decade or two, it has been common on the eastern coast of the United States. This is possibly both an infection and a poisoning, since not only is the abdominal pain accompanied by nausea, vomiting, and diarrhea, but also blood and mucous are found in the stools in much the same manner as in the severe dysentery infections.

There are many diseases transmitted by foodstuffs and those that have historically been of importance will not be discussed here. At one time, diseases such as cholera, diphtheria, scarlet fever, and many others were transmitted through foods, including milk. Through the ages, the most dramatic has been typhoid fever. Latent carriers, not themselves ill, brought sickness and even death to tens of thousands. The name "Typhoid Mary" came to be applied to carriers of all types of diseases. With the great reduction of typhoid disease throughout the world, this is not of as great concern at present. There have also been infestations produced by lower forms of life such as flatworms and flukes, often through external water contact.

The important reportable diseases that are foodborne, and that can also be waterborne in some cases, are the ones caused by bacteria of which *Salmonellosis*, with approximately 28,000 cases reported per year, and *Shigellosis*, with 16,000 cases reported per year, are by far the most important. One protozoan form causes amebic dysentery, and 3000 cases are reported each year. Since *Staphylococcus* food poisoning and *Cl. perfringens* are not reported officially, we can only estimate that there are probably 500,000 cases of each annually, similar to the actual incidence of Salmonellosis.

Almost without exception, everyone who reads this book will have had salmonellosis at some time or the other. Rarely was a physician consulted. Within ten to thirty-six hours after eating a particular foodstuff, you probably developed diarrhea. If the case was more severe, you may have had nausea, vomiting, fever and a headache. Most of the symptoms disappeared within one to two days so you said to yourself, "It's something I ate"—which was certainly true. The physician usually only sees the severe cases of greater abdominal distress, aching of the limbs, and prolonged diarrhea lasting three to eight days. The symptoms of the more mild *Shigella* cases will be similar.

In each instance, usually a human carrier of a certain strain of organism is responsible. There are approximately 2000 serotypes of *Salmonella*, and, in some instances, waitresses or cooks carrying three different strains have been found. For each of the last ten years, the deaths have varied between about sixty and eighty per year in the United States. While this is low in comparison to the number of reported cases, a death rate of 0.3 percent is much too high in a disease that is preventable.

Outbreaks have originated from raw and raw certified milk, from spray-dried milk and eggs, and from meat products in which the organisms have been present in the original food. If milk that is not pasteurized is permitted to stand at ambient temperature for several hours, the number of *Salmonella* can cause illness, particularly to an infant.

Much of the market poultry contains *Salmonella*. With chickens, certain surveys have shown that over half of these have the organisms present. There is little danger from eating fried chicken. Most of the problem arises when chickens or turkeys, particularly the latter, are stuffed with a dressing containing the chopped giblets. These will have *Salmonella* if the original poultry was infected. Usually, in roasting a turkey, heat penetration is not sufficient to destroy the organism in the middle of the dressing. For this reason, if at holiday seasons the turkey is baked the night before, the dressing should be removed, and both should be refrigerated, for the cold will not penetrate to the center. Otherwise, the organisms can grow to be thousands per gram of dressing. The next day, the dressing can be heated separately from the bird and then restuffed prior to bringing the bird to the table.

If one is preparing chicken salad, the raw poultry should never be cut up on the same board that will be used to dice the cooked meat unless the board is of hard plastic and thoroughly washed and disinfected for use of the cooked meat. Microorganisms cannot be removed from a wooden cutting board, so hard plastic is far better. Every hotel and hospital should have two sets of cutting boards, one for raw meats and the other for cooked meats. I have known of one outbreak resulting from the use of chicken salad at a church supper where the cooked, diced meat had been handled as above, then permitted to stand at ambient temperature during which time *Salmonella* increased in numbers and most of the persons became ill.

The same danger lies in convenience foods such as TV dinners and one-dish meals. These should always be heated for the full length of time stated on the package, as this schedule has been rather carefully worked out to destroy harmful bacteria.

Shigellosis and enteropathogenic *E. coli* produce infections similar to salmonellosis and for this reason will not be discussed separately. The deaths from *Shigella* infections closely approximate those from Salmonellosis.

There are several salmonellae that cause enteric fevers. These are entirely different from the salmonelloses which resemble food poisoning with the quick onset of symptoms. Typhoid fever, of which close to 400 cases are reported each year, and the paratyphoids which are not reported, have incubation periods of seven to ten days. In typhoid fever, the bloodstream is invaded and a whole chain of symptoms associated with the fever develop. The death rate at present is a fraction of 1 percent, with the advent of "miracle drugs."

Personal cleanliness in thoroughly washing hands after using the toilet, thorough cooking, and adequate refrigeration and freezing of foods assist in removing foods as a source of great danger.

The UN Children's Fund, working with the World Health Organization, stated that at least one million children die annually because of inadequate bottle feeding in the Third World nations. The difficulty is that the giant food companies, of which Nestle's is an example, have supersalesmen pushing nourishing dry-food formulations. When liquified with water and placed in nonsterile bottles at ambient temperatures, the previously mentioned microorganisms can cause illness and death. I am ashamed that my country, in May 1981, was the one country that voted when the vote was 95 to 1 *not* to restrict the marketing of these infant formulations.

In dealing with amebiasis or amebic dysentery the same principles apply. Throughout history, more battles—and even wars—have been lost to dysentery than to enemy action. At present, of the approximately 3000 cases reported caused by the protozoan parasite *Endamoeba histolytica*, many of these have been acquired abroad, chiefly in Mexico. Other cases are due to the food handler being a carrier of the organism. This parasite passes from the intestinal tract into water through fecal discharges. Many towns and cities in Mexico do not have safe water supplies. Even in Mexico City this is true. The capital is a sinking city, sewage pipes burst, water pipes strain at the joints, and the latter can siphon contaminated water. It is best not to use green salads unless one is positive the vegetables have not been washed with contaminated water. If possible, one should drink water he has chlorinated and dechlorinated himself. It is not inconvenient to carry along a quart container which has a built-in heating unit. Boiled water should even be used for brushing one's teeth.

Amebic dysentery is a painful disease, and, if not treated adequately, the organisms can invade the liver or even the brain. When traveling abroad, take care. Stay with well-heated foods, if possible.

Tularemia will not be discussed since this is a disease of hunters who contract it from the skins of diseased rabbits or squirrels; it is not a true foodborne infection. Almost none of the cases of tuberculosis are foodborne since, in these days, cattle are tested and milk is nearly always pasteurized. There are still from 100 to 200 cases of trichinosis annually caused by eating inadequately cooked pork, or, occasionally, game animals such as bears. Foodborne virus diseases do occur but these organisms do not multiply in foods; instead, the latter serve to carry the virus from a carrier to a victim.

6. Poisonous non-foods. To this point, we've been discussing foods and beverages intentionally ingested for nutritional purposes, but which—either by reason of personal idiosyncracies or by nature of that food—can be harmful even if not contaminated. They still, however, remain in the category of foods. We've also seen how external contamination, or "tainting" by disease mechanisms, can transform them into quick-striking monsters.

There are, however, other substances—usually, but not always, flora—which are not foods and never should be mistaken for edible substances. Immediately coming to mind, I am sure, will be toadstools. Here, again, "a little learning is a dangerous thing," as Pope reminds us, and it is usually the person who knows only a little about mushrooms who gets into trouble.

One mycologist (mushroom expert) stated that he wouldn't presume to go out to a wooded area and attempt to select safe varieties. "There is so much mimicry in nature," he said, "that it would be hazardous even for me."

Those who pick mushrooms for their family table usually limit such selection to a single variety with which they have become familiar from childhood. The two most dangerous types—sometimes used in fictional plots—are *Amanita phalloides* (death cup) and *Amanita verna* (destroying angel). Mushrooms retailed commercially by well-known food processors are quite edible and much safer.

Water hemlock, *Cicuta maculata*, which is also known as wild parsnip, has caused illness and death in children. This plant thrives in the vicinity of irrigation ditches and swampy areas.

White snake-root, also known as jimmy weed, belongs to the genus *Eupatorium* and can cause illness and death in the human. Cows in woodland pastures in which this plant is present may eat it. The compound tremetol is formed, and as a result, can be present in large amounts in the milk of the animal. In pioneer times this poisoning was known as trembles, and is said to have caused the death of the mother of Abraham Lincoln. Keats mentioned, and one may include:

> No, no, go not to Lethe, neither twist
> Wolf's-bane, tight rooted, for its poisonous wine.

The castor bean plant which grows wild, among other places, on Mount Washington, contains poisons, one of which is ricin. Children have eaten castor beans with resulting illness or death. Deadly nightshade, *Atropa belladonna*, contains the poisons atropine, hyoscyamine, and scopalomine. Three or four of the berries can be toxic to children who eat them, mistaking them for an edible berry.

Conium maculatum, or hemlock, has caused death when the leaves have been used instead of parsley. The root of this plant also has been mistaken for wild parsnip, causing poisoning since it also contains toxic alkaloids including coniine. There are many other examples, and parents who have small children should teach them to avoid chewing upon or otherwise using plants unless they know definitely they are nontoxic. Even the oleander which grows as a decorative shrub in many California lawns contains toxic materials. Frequently, children camping out have cut oleander twigs and used them to roast wieners, causing severe illness. The leaves and twigs of many evergreen bushes and shrubs are also poisonous.

Vitamins such as Vitamin A, the B vitamin group, C, D, E, K, and B_{12} are considered essential for man. Damage can be produced by continued large doses of vitamin A and D. Large amounts of nicotinic acid produces flushing of the skin, cramps, headache, and nausea. Folic acid is not needed at more than 0.1 milligram per day and massive doses can result in renal damage.

Factors that can be most damaging, particularly to animals, are the antivitamins and antimetabolites. These are substances that destroy or bind up the vitamins so they cannot be used.

Raw carp, certain other fish, and crustacea contain thiaminase, which destroys the enzyme thiamin. It was found on mink and fox farms that the fish had to be heated to destroy the thiaminase or fed at a different meal. The bracken fern has an antithiamin activity that caused difficulty in animals grazing in poor pastures in some areas of the world.

It has already been mentioned, in the last chapter, that niacin is bound tightly in corn and that treatment with dilute lye would release it in its effective form. Without this, consumers can develop pellegra.

Avidin in raw egg white binds the needed vitamin biotin, and conalbumin, another substance present, binds iron needed by the body. These binders can be inactivated by heating.

The oxalic acid present in rhubarb or spinach binds calcium, and, as a result, it is necessary to add an increased amount of calcium to the diet. Vitamin K is a clotting factor of the blood and can be destroyed by the dicoumarol in sweet clover.

There are factors in soybeans that cause thyroid enlargement and others that interfere with protein digestion. These are only a few of the dozens of conditions that exist in certain foods, but are the ones of most concern.

7. Summary. It is important to point out dangers in foods— dangers which become more intensified as populations increase in density. But we should not become so alarmed over "possibilities" that we lose sight of how far we have traveled in the improvement of food safety. As present, we enjoy a greater variety of available healthful foods than ever before in history. If we do not select from them a proper balance of nourishing items, such cannot be blamed upon food processors or merchants. We would be better off, physically, to reduce our intakes of salt, sugar, fats,* and perhaps protein—at least in red meat form. If we then use high standards of cleanliness, and prepare our foods properly, we can truthfully say, "Grandpa never had it so good!"

*Often butter fats are reduced in amount, with the label stating, "vegetable fats added." This frequently is coconut oil, which is a highly saturated fat and probably less desirable than the butter fat it replaces.

The Cides That Kill or Injure Us*

In a later chapter** we will be discussing how man developed from a grubber for roots and insects up to our present thickly populated urban communities. In prehistoric times man probably was not bothered by plagues of grasshoppers and locusts for if or when these occurred, they became just another component of his diet. It was not until man had domesticated certain animals and had begun to cultivate crops that the insects and arthropods*** that preyed upon his grain crops became competitors for his foodstuffs. At this time he searched for methods of destroying these pests. Weeds also became a problem, but until recently, he destroyed these by pulling them or by the process of cultivation.

Also, in these most primitive times disease was probably a problem of the individual or family, or in certain rare instances of a clan. As city-states grew and the population was crammed closely together within

---

*The term "cides" refers chiefly to biocides and pesticides, but embraces the gamut of such terms.

**See "Our Exploding Populations," page 170.

***Arthropods have segmented bodies and jointed limbs, such as grasshoppers.

walls, other problems developed. Armies of tens of thousands marched to conquer other countries or empires. Soldiers were crowded under unsanitary conditions. The great plagues of mankind, in some of which insect or arthropod vectors were involved, swept through these populations at peace or at war. Throughout history more armies were destroyed by disease than by battle. It was only recently that through the use of pesticides this danger has been reversed.

With this in mind, let us discuss how the insect and arthropod vectors became important in agriculture and diseases of the human up to the present time. In the future we hope to use organisms of diseases that attack the specific pests, or hormones extracted from their own bodies, to destroy them. Such methodology has begun for some pests.

In this long process many pesticides were used and discarded, and methods of control other than pesticides were employed. In the struggle to destroy the weeds that damaged crops, certain chemicals were developed, a number of which are so deadly that they, as well as some pesticides, became weapons of warfare. Now mankind is groping at the threshold of new knowledge in which advanced biological methods will be used.

With this in mind, let us roll back the curtain some thousands of years and briefly examine the situation up to the present.

Insect pests and plant diseases were known from the time that man planted his crops in relatively large fields and particularly where the same crops were planted year after year. This permitted insects to multiply and remultiply under ideal conditions. The prayer of Solomon at the dedication of the temple illustrates this clearly:

> Then hear thou in heaven, and forgive the sin of thy servants, and of thy people Israel, that thou teach them the good way wherein they should walk, and give rain upon thy land, which thou hast given to thy people as inheritance. If there be in the land famine, if there be pestilence, blasting, mildew, locust, or if there be caterpillar.

In these early times, if grasshoppers or locusts descended upon the land they could only be fought by beating at them with flails, boards, or sacks. Possibly fires were used at times to stop their advance, as has been done up until relatively modern times.

In 1000 B.C. Homer mentions "pest-averting sulfur." It is known that if soil is sulfured, certain bacteria will oxidize this to sulfur dioxide and sulfuric acid. This treatment has been used in some vineyards for centuries. Sulfuring soil prevents a disease of potatoes caused by *Streptomyces*, since the active species are not pathogenic in acid soils.

Democritus in the fifth century B.C. spoke of the use of amurea of olives for blight control. Theophratus in the fourth century B.C. believed there were a number of causes for crop failure and his description is almost that of a modern agronomist, as he includes unfavorable weather, unfavorable soil, plant diseases, and finally, the will of the gods. Shakespeare in *King Lear* recognizes the foul fiend Flibberty Gibbet who mildews the white wheat.

For almost countless centuries the poor of Europe lived upon rye as their primary foodstuff. During years of inclement weather the ergot fungus invaded the grain, causing disease. At one time this was known as Saint Anthony's Fire due to the symptoms of the fingers and toes becoming black and gangrenous before the death of the individual. (See Chapter 5.) Within the last three decades there have been unfavorable seasons in which this fungus produced ergotism and death in the people of France. Now that the cause is known, severe penalties, including the death penalty, can be imposed upon millers in certain countries who deliberately use contaminated grains.

Wheat rust was one of the severe plant diseases of Europe, and in the 1600's it was recognized that the common barberry was the alternate host of the disease. In France at Rouen in 1660 the common barberry was eradicated. In the 1700's many of the American colonies followed a similar eradication program. In 1881 there was a quarantine on the United States because of an infection occurring in grapes. Since this time, many states have set up quarantines so that fruit, vegetables, and flowers cannot be brought into those states because they might introduce parasites. An individual who tries to evade these quarantine regulations is willfully reckless since he can, thereby, introduce new parasites that can take millions of dollars to control.

Insofar as pesticides are concerned they are complex, but this subject will be simplified. They belong in three groups called first generation, second generation, and third generation pesticides. Before discussing these remedies, it might be wise to look at certain diseases spread by insects and arthropods. As Thomas Nash said, "From winter, plague and pestilence, good Lord, deliver us."

Going back into early Biblical history, we are quite certain from the writings of the Assyrians that the great army of Sennacherib besieging Jerusalem was destroyed by typhus fever. This disease in epidemic form is spread from man to man by means of the body louse. This disease struck again and again at armies in the field and sometimes at civilian populations. It destroyed the Grand Army of Napoleon, and, although he captured Moscow, his men were dying by the tens of thousands and relatively few lived to retreat back so far as Germany. Near the end of World War I, typhus fever started among the Russian troops and was carried into the civilian population, killing millions. In World War II, when the Allied armies invaded Sicily, typhus fever was just starting among the civilian population which had inadequate food, clothing, or cleanliness. Fortunately, at this time a new second generation pesticide called DDT had been developed. The civilian population was dusted and there were few cases of typhus.

Another great killer of man that could not be fought until recently was plague, also known as the Black Death. It had its maximum impact as the city-states grew in population and size. In A.D. 70, when Rome had a population of over a million persons, this disease, spread first from the rodent fleas to man, caused a massive epidemic. At its height, 10,000 persons died per day.

A pandemic, which is a worldwide epidemic, began in lower Egypt in A.D. 542 during the reign of Emperor Justinian. It lasted for half of a century, killing untold numbers of persons. In the middle of the fourteenth century another pandemic of plague swept through Europe, destroying two-thirds of the population. This great plague occurred between the times of the battles of Crecy and Poitiers between England and France, making the great death rates inflicted by battle seem miniscule.

Plague again struck Europe during the reign of Charles II of England. An excellent description can be obtained from either Samuel Pepy's *Diary* or Daniel Defoe's *Journal of the Plague Year*, the latter being written as an account learned from persons who had experienced its ravages. Intermittent cases of plague occur throughout California and, within the past week as I write this, one person has returned home cured. Now with our second generation pesticides we have means of destroying rodent fleas which distribute this disease.

To go back for a moment, first generation pesticides included a number of compounds that were relatively effective for different purposes. Volatile compounds such as carbon disulfide were worked into the soil around the roots of grapevines to destroy microorganisms, nematodes, or higher forms of life that could injure these plants.

Copper arsenate, copper sulfate also known as Bordeaux mixture, and lead arsenate were some of the compounds used to destroy pests. These were nonspecific poisons in that the heavy metal reacted with the protein of the microorganisms, insects, arthropods, or higher forms of life such as caterpillars or tomato hornworms. Chewing insects were poisoned as the materials entered the intestinal tract and were distributed throughout the body. People also were poisoned if they ate the food that contained large amounts of these pesticides.

Other first generation pesticides included mercury-containing compounds that were efficient fungicides and could be used for seed treatment to prevent "damping off," which is the death of the sprouting seed. Nicotine solutions were effective sprays to combat aphids and other sucking insects. A few similar sprays or dusts made from natural compounds include rotenone and the pyrethrins, which are extracted from plants.

Kerosene and other oils are also first generation pesticides and are used to kill the larvae of the Anopheles mosquito that carries the parasite causing malaria. This is done by pouring a thin film of oil on ponds, ditches, or other water sources where the mosquito is breeding. The oil film shuts off the oxygen of the air from the "wigglers," as the larvae are sometimes called, and as a result they die. The same method has been used to destroy breeding areas of mosquitos that carry the organisms of the disease known as elephantiasis.

I am making a list of the three groups of second generation pesticides since the names are commonly encountered in newspapers and magazines. Although each group contains more names than the following lists, these are sufficient to include the more usual.

Type of Pesticide	Common Examples	
Chlorinated hydrocarbons (Non-biodegradable; lasts for years in soil or water)	DDT DDD aldrin chlordane dieldrin	heptachlor Endrin lindane toxaphene endosulfan
Organophosphates or "Nerve gases" (Biodegradable; disappears in days or weeks in soil or water)	parathion malathion diazinon	
Carbamates	Sevin	

The chlorinated hydrocarbons are not specific poisons of one group of nerves or hormones. The danger of superchlorination of water lies in the fact that some of these chlorinated hydrocarbons will be produced and possibly be carcinogenic in man. Most of these pesticides have been important. Some, however, are losing their efficiency as insects adapt to them, and others can be dangerous to lower animal forms and to man.

I have no intention of debating the sociopolitical effects of any of the second generation pesticides. However, in India in 1955 there were approximately 250 million cases of malaria with a debility and death rate of 20 percent. DDT was used against the *Anopheles* mosquito with the result that in 1962 the number of cases had decreased to 140 million, and in 1967 to ten million cases. Despite the fact that the mosquito is developing resistance to DDT and certain other biocides, the number of cases of malaria declined to about three million in 1970. In 1973, the number of cases of malaria increased slightly to 3.9 million, due both to the increasing resistance of the mosquito to DDT and the fact that the price of DDT had increased from \$475 to \$1500 per ton. Where mosquito resistance is high, Dr. Sadanand Pathnayak stated that malathion also is employed for control. In India a remarkable decline in typhus fever also occurred due to the use of DDT and its ability to kill body lice. A few economists have wondered whether this has been beneficial to India since, before the use of biocides and immunization for certain other diseases, the death rate was such that population control was assured. This is no longer true.

The secondary generation pesticides, being nonspecific, have a shotgun effect upon insects. One of the best illustrations of this concerns disease and the citrus crops of California. In the early 1880's our citrus crops were attacked by cottony cushion scale. A natural enemy of the parasite, the *Vedalia* commonly known as the ladybug, was introduced from Australia and within a few years the scale had almost disappeared. In the late 1940's and 1950's DDT was used to destroy other parasites; and, since it is nonspecific, it destroyed the *Vedalia*. Cottony cushion scale, with its damaging effects, again became an important pest. Fortunately, California has stopped using DDT in such orchards and again has been able to reintroduce the *Vedalia* to normalize the situation.

Pests such as the cotton boll weevil, the corn earworm moth, the sugarcane borer, and the rice water weevil, that at one time were controlled with one application of relatively small amounts of this insecticide and its close relatives, now require several applications of larger amounts.

During the months of May and June of 1979, Madera County, California, was invaded by grasshoppers that covered and stripped lawns, home gardens, and flowerbeds. Almost simultaneously, Yakima Valley in Washington was also subjected to attack and up to thirty young grasshoppers were found per square yard. In the Los Angeles *Times* of June 6, 1979, there was a report from South Dakota State Entomologist Ben Kantack that 800,000 to 1,000,000 acres of that state were loaded with grasshoppers. He stated that there were as many as 100 per square yard and also commented that eight grasshoppers per square yard eat as much grass as a cow. Ranchers were afraid that the government spraying programs would be too late to save their crops. As Edward Burke said:

> Because half-a-dozen grasshoppers under a fern make the field ring with their importunate chink, whilst thousands of great cattle, reposed beneath the shadow of the British oak, chew the cud and are silent, pray do not imagine that those who make the noise are the only inhabitants of the field; that, of course, they are many in number; or that, after all, they are other than the little shriveled, meager, hopping, though loud and troublesome insects of the hour.

In March of 1979 a plague of locusts in Kenya, East Africa, was barely being controlled. The swarms were first spotted in the Jizan region of Saudi Arabia in 1978, and as the months progressed, the swarms of millions moved southward into Ethiopia, Somalia, and Djibouti, Kenya, which they had reached in June 1978. In Somalia alone almost 100 swarms were destroyed. Control was difficult in some areas because in the Ogaden area claimed by Ethiopia and Somalia, the fighting made great areas inaccessible. The same was true in the Eritrea province of Ethiopia.

Although DDT and most other pesticides can be used in Africa, India, and many other areas of the world, its use is banned in the United States by the Environmental Protection Agency. Even under the present emergency situation of ground squirrels and other rodents harboring the rodent flea that carries plague mentioned previously as the Black Death, I am not certain if such permission has been granted.

On May 22, 1979, it was reported that Maine planned to spray four different insecticides over 3.25 million acres of timberland to protect it from the spruce budworm, which is destroying the timber. Environmentalists recently had brought suit against the state to block the program as being harmful to health and not economically sound. This is a difficult situation since much of the timberland of the northeastern and northwestern portions of the United States has been and is being destroyed. I believe that third generation pesticides, which will be discussed later, will eventually replace most of these second generation pesticides—but until that time, limited, carefully controlled spraying must be permitted.

The Maine Forest Service has, however, dropped plans to spray 20,000 acres of spruce budworm-infested forest with Metacil, which is claimed to be a viral enhancer. Viruses normally dormant in the host become active and may destroy it. Much laboratory testing must precede its use to be certain that this does not activate viruses in other animals or in humans. This is an example of a third generation pesticide, which was inserted here to indicate that all is not perfect in any class.

From March through May 1979, and perhaps longer, the residents of Triana, a community near Huntsville, Alabama, were being tested to determine the blood levels of DDT on those persons who have eaten fish from the local streams. The National Center for Disease Control stated that a number of persons have blood levels two or three times higher than other population groups. The source of the pesticide was from the site of a former DDT plant operated on Army property at Redstone Arsenal for approximately thirty years by the Olin Chemical Corporation, according to the Los Angeles *Times* of May 14, 1979. The plant itself was closed down in the early 1970's.

In May 1979, the U. S. House Agriculture Committee voted to allow the use of the pesticide Mirex, which causes cancer and birth defects in laboratory animals, against fire ants for an additional year and simultaneously limited EPA regulation of pesticides. The EPA immediately promised to seek a veto from the president if this passed Congress, as Douglas Costle stated that Mirex may be a cancer-causing agent. Instead, the EPA recommends the use of diazinon and imidan, two agents that are less harmful. Ferriamicide is another pesticide used against fire ants that might be dangerous to man.

The EPA in June 1979 has permitted pear growers to use the pesticide Baam on pears if the workers wear protective clothing and pears are not harvested until seven days after the pesticide has been applied. This is another pesticide that is suspect and all purchasers of fruit should scrub fruits, when possible, before eating.

California still rejects the use of chlordimeform as a cotton pesticide, as the state believes the workers cannot be protected adequately from this carcinogenic substance.

A tremendous argument has raged over the safety and use of DBCP (Dibromochloropropane) after it was discovered that certain workers formulating the product in the Occidental Chemical Company plant were either sterile or had low sperm counts. The material is so dangerous to health that no more than one part per billion can escape into the work space within a chemical plant. This means that one single drop in a room twenty by twenty feet would exceed these standards by ten times, according to Stephen Rappaport, a University of California public health expert, in January 1979. The one plant manufacturing this highly toxic compound is Amvac, located in Los Angeles. On either December 18 or December 20, 1978, according to employee versus employer reports, fifteen to twenty gallons of this material were spilled onto the concrete loading dock and the employees had to clean this up without adequate protective clothing or respirators. Richard Wade, Deputy Chief of CDOSH (California Department of Safety and Health), announced that Amvac was permitting 500 parts per billion to escape from their stacks into the air. I am particularly intrigued by the escapes involving this highly toxic substance for killing nematodes since, for some time, I was working in a consulting office not too far from this plant. It is interesting to note that in October 1979, EPA Administrator Douglas M. Costle said there was substantial evidence that DBCP could cause cancer or sterility in workers and banned the use in every state except Hawaii. This exception was over the objection of his own staff members.

It is also ironic that DBCP was developed by the Pineapple Research Institute in Hawaii more than thirty years ago to kill nematodes in the soil. In May 1980, however, DBCP had been found in wells on Oahu, among which was one well serving the tiny Del Monte town of Kunai. Undoubtedly, its use will be barred* and another compound lethal to nematodes substituted. We hope the situation is not what Thomas Nash would describe:

> Brightness falls from the air;
> Queens have died young and fair;
> Dust has closed Helen's eyes.
> I am sick, I must die.
> Lord, have mercy on us!

*Still in use in May 1981 in Hawaii.

Another pesticide that has caused great difficulty is Toxaphene. This is used primarily for spraying cattle to eliminate psoroptic cattle scabies. A series of articles in the Los Angeles *Times* from February 2, 1979, through March 16, 1979, concern this pesticide. Rancher George Neary claimed the spraying by the California Department of Food and Agriculture killed 250 cattle on his ranch. Assistant department director Herbert Mills confirmed that the departmental employees violated regulations by allowing some of the material to drain upon the ground. Jim Pedri, an officer of the State Water Control Board at Redding, confirmed an additional charge that excess Toxaphene ran into a nearby creek and may have contaminated the Sacramento River. In September, 1979, Federal investigators concluded that state veterinarians' misuse of Toxaphene had contributed to the deaths and abortions involving over 500 cattle on the George Neary ranch.

For almost a year and a half, a group of people worked at Life Science Products Company producing a pesticide called Kepone for Allied Chemical Corporation. The workers stated they were not warned that the product was dangerous. A series of illnesses in July 1975 closed the plant when seventy persons had been poisoned with Kepone, twenty-nine of whom were hospitalized. The symptoms included liver damage, tremors, and, in one case, sterility. Vast areas of the James River had been contaminated with the effluent and this rich seafood area was declared off-limits. Allied Chemical pleaded no contest to 940 misdemeanor charges of pollution. Heavy fines into the millions were assessed.

A bill to make it harder for pesticides to qualify for use in California passed the Assembly on May 29, 1979, and moved to the Senate where it was approved and adopted. The California Department of Food and Agriculture is self-critical and is working hard to determine more closely the toxicity of the compounds used and to improve the chemical testing for residues.

Another problem that occurs in probably every state is that crop dusters or sprayers miss or overshoot their fields and spray workers in adjacent areas.* This is particularly dangerous when pesticides that are nerve gases, such as parathion, are used. These will be discussed briefly after some of the dangers of the halogenated hydrocarbons have been illustrated.

*In 1981 it was noted that the pesticide carbaryl drifted as far as eighty miles from the site of spraying timberland for spruce budworms near Augusta, Maine.

It has been known that DDT and most of the halogenated hydro-carbons can persist in the soil for months or even years. As a result, when an orchard or forest is sprayed with DDT, sufficient amounts are carried through the air so that eight to nine milligrams per cubic meter of air can be found as far away as a mile. Naturally, the major amount lands upon the trees, or the trees and soil, if the area is sprayed by airplane. Hopefully, the pests involved are destroyed. The great difficulty, however, lies in what we call the biological magnification of these pesticides. One very simple example exists in that, even were the soil in the orchard not contaminated in the dusting, with each rainfall more and more DDT will be washed to the soil. Earthworms and burrowing insects become grossly contaminated by the product. Robins, red-winged blackbirds, and other birds that eat the insect pests are destroyed either directly or because they cannot reproduce. The seed-eating birds such as crows and starlings which damage fields are not injured.

Although the DDT is practically insoluble in water, it is washed into the streams and carried to the rivers, lakes, and ocean, where it is extremely damaging because of the biological magnification. These halogenated hydrocarbons can dissolve in the fat of algae or other phytoplankton. If a sufficient amount is stored, photosynthesis by the algae can be interrupted. The most important point, however, is that the zooplankton feeding upon these tiny plants concentrate the DDT in their bodies, and, when larger forms feed upon these, great concentrations occur. For example, the oyster and other shellfish can concentrate the halogenated hydrocarbons up to at least 1000 times that of the surrounding area. The mature oyster passes approximately sixteen liters, or about the same number of quarts, of water an hour through its gill system in extracting the plankton, and the oyster feeds about 90 percent of the time.

A rather similar situation occurs with small fish eating zooplankton or insects. When Clear Lake in California was sprayed to destroy midges, although the water showed negligible amounts of DDT, 200 parts per million were discovered in fish flesh and 2500 parts per million in the visceral fat of the fish, which had concentrated the DDT from the insects they had eaten. This is only one example, many of which have occurred in using long-lasting pesticides. Fish-eating birds can concentrate DDT still further.

It has been noted since the 1950's that certain predatory birds were disappearing. Ratcliffe in Britain noted that the peregrine falcon, which has been a popular bird since the Middle Ages, was becoming extinct. Risebrough at Berkeley noted that the pelicans on Anacapa Island were becoming fewer. A host of other observers found the same to be true for the eagle, osprey, kestrel, Cooper's hawk, and other birds. They all had a point in common in that these species ate larger and larger fish, or, in some instances, land animals in which the concentrations of DDT were greatly magnified. Risebrough noted that DDE, a breakdown product of DDT, could be as high as 2500 parts per million in the eggs and that relatively small amounts such as 75 ppm reduced the shell thickness by 20 percent. This is consistent with the work of Ratcliffe.

The thinness of the shell is due to the fact that in birds the calcium of the shell is laid down in the last twenty hours, a longer period with certain larger birds. The calcium of the eggshell is drawn in part from the food supply of the bird, in part from the calcium reserves of the bones of the bird; in each case being carried by the blood stream. An enzyme known as carbonic anhydrase makes the supply of calcium carried to the oviduct available. This enzyme is blocked by the halogenated hydrocarbons such as DDE. Accordingly, the eggs when laid have shells so thin they break readily and the young cannot mature.

Delayed breeding is another factor in the decline of the population of these birds. The halogenated hydrocarbons affect the hormonal levels of the birds so that mating at the appropriate time does not occur.

It should be noted that the compounds known as polycholorinated biphenyls (PCB) act in the same way as do the preceding compounds, but exert a stronger effect. The PCB's are used as insulation compounds in electrical transformers and their use and disposal should be carefully controlled.

In May 1981 a Southern California Edison Company capacitor exploded, spraying the home, garden, and swing set of a family with toxic PCB. An Edison Company official stated that such accidents are not uncommon. PCB's have been found in hospitals from leaking transformers, in gas lines in Long Island, Chicago, and Los Angeles, and in many areas where this carcinogen has been dumped or inadequately stored. It is ironic that the New York State Department of

Environmental Conservation had their own building coated inside and out when an electrical transformer containing PCB burst during a fire, resulting in ash contamination of the building. The best solution to the problem would be to fill the transformers with jojoba oil or preferably with a cheaper nontoxic material.

Before leaving the area of halogenated hydrocarbons, I should mention that in the studies made in past years, cow's milk had a lower content of DDT and DDE than did mother's milk, as shown by one article in the *British Medical Journal* in the mid-1960's. This would have been true in areas such as California where alfalfa for cattle feeding had to be tested before sale. It would probably not be true at the present time, although certain other pesticides might be high due to the use of spray disinfectants for roaches, ants, and other pests in the home. Richard Purdy Wilbur might have said of the toxic substances in milks:

> We milk the cow of the world, and as we do
> We whisper in her ear, "You are not true."

Body fats excised from the human and examined for DDT a decade ago were almost in exact ratio to the amount of pesticides used in the countries concerned. It is probable that the ratios presently found in Hungary, India, and Israel have continued at such a high level because those nations continue to use DDT. The percentage of halogenated hydrocarbons in fat has dropped in the United States and Western Europe, where DDT has been banned.

The other main group of second generation pesticides that have been mentioned, but not discussed, are the organic phosphorus compounds which were originated as nerve gases but not used in World War II. There are more than a dozen of these, but the ones most commonly encountered in study are parathion, methyl parathion, and malathion.

These compounds are much more toxic than the halogenated hydrocarbons. A drop of pure parathion on the skin or the inhalation of a small amount of the vapor would be toxic quite rapidly. Although these are used in crop dusting, it should be realized that as either spray or powder they are normally diluted. There have been accidents in California in which a crop duster has overshot his mark and sprayed some of the laborers working in adjacent fields. Illness has resulted but, insofar as I remember, all of the persons were hospitalized immediately and death did not occur. I do remember one instance in Mexico in which a number of drums of parathion were stored in a government warehouse where a number of other materials, including flour, were also stored. One of the drums leaked, contaminating the flour, which was later released to a local baker. His products caused illness and death among a number of his customers. He was jailed and might have been tried and given a long jail sentence, or even executed, when the government found, to their embarrassment, that the material had been contaminated in the government warehouse.

In October 1979, state officials halted the harvest of tomatoes in North San Diego County when they discovered growers had violated both state and federal laws by spraying the crop with Orthene. Over twenty tons were dumped and almost a hundred tons withheld from shipment. In March and April 1980, border inspectors discovered that Mexican growers had been using the highly toxic German pesticide Celathion-50, which is toxic to the fetus and damages human nervous systems. On the farms in Mexico, the growers gave no instruction about the toxicity of the chemical, and the empty pesticide containers were distributed and used to hold water for washing or drinking.

Organophosphorus compounds do not belong in the home but they have certain advantages in agricultural use. These compounds are either spontaneously or biologically biodegradable. When used with moderation, they disappear within three weeks to four weeks after spraying and do not remain as contaminants for years.

The carbamate compounds, of which probably Sevin, Zectran, and Baygon are the most commonly used, also are disintegrated within three to four weeks. It should be remembered, however, that the six or eight compounds within this group have been opposed because of the effects they might have upon humans, particularly women during pregnancy.

Third generation pesticides are those which kill only pests and have no effect whatsoever upon humans. These fall within several groups, and although they have been used successfully for the control of a few pests, we are just pushing the doors open through which we can eventually destroy many. I would group these into the following: juvenile hormones, insect attractants, sterilization or partial sterilization by using radioactive cobalt or x-ray, bacterial and virus diseases of pests, and, last, the breeding of predatory forms.

Juvenile hormones are present in insects that go through meta-morphosis, that is, through a number of forms before they become a mature adult. The final compound, ecdysone, initiates the molts through which the insect must pass. It must be absent from the egg, present in the developing larva and in the pupa, and absent in the later stages of molt. In natural Cecropia moths and certain other forms, this could be successfully used if the juvenile hormones could be synthe-sized in the laboratory in sufficient amounts to be used as sprays. To date, limited success has occurred.

Certain pests, particularly the female, produce compounds known as pheromones, or sexual attractants. There are a number of pests susceptible to these, but to secure the pure compound to attract the silkworm moth, it is necessary to use the abdominal tips of 500,000 virgin* females to secure approximately 1/2000 of an ounce. Com-pounds of this sort can be used to attract the Mediterranean fruit fly, the gypsy moth, and other pests into traps.

Sterilization or partial sterilization of an insect has been used to suppress certain pests. I have made up a complete formula for the cattle screwworm fly (*Cochliomyia hominivorax*), upon which the fly could lay its eggs. These hatch and go through the metamorphic stages. Just before the development of the mature fly, cylinders loaded with tens of thousands of the mature male pupae are exposed to radioactive cobalt and the insects sterilized. The sterile flies are released in infested areas so that there are approximately ten sterile flies to each wild type. Since the fly breeds only once, there is little chance of a fertile mating. Programs similar to this, in which I have not had a part, have been developed which include the Mediterranean fruit fly. California has been torn between advocates of the use of pesticides and the sterile fly technique during 1980 and 1981. While the sterile fly program worked in Florida, to this point it has not worked in California; the authorities started too late. Similar types of programs can be achieved by the release of partially sterile male Lepidoptera such as the codling moth, cabbage looper, pink bollworm, and tobacco budworm.

Bacteria and their toxic products such as *Bacillus thurengensis* have been used to destroy insect pests such as the alfalfa caterpillar, the cabbage worm, cabbage looper, tomato hornworm, and tobacco bud-worm. This organism is completely harmless to the human but kills

---

*Virginity, as may be seen, is not always a blessing.

chewing insects by paralyzing their stomachs. Several years ago a variant of this organism, *Bacillus thurengensis I* (Israeli), was discovered, to which some pests that were resistant to the main strain now succumb. It is now being used extensively against mosquito larvae, since it does not harm most insects which feed upon mosquitos.

Bacillus popilleae invaded the Japanese beetle, producing milky spore disease, and killing this pest. Viruses have been used against the beet army worm, cabbage looper, codling moth, corn earworm, cotton leaf worm, full army worm, red-banded leaf roller, European spruce budworm, tobacco budworm, pink bollworm, and alfalfa looper. Although more work needs to be done with the above, and with the other viruses and bacteria that will kill pests, it is obvious that if we succeed we will have eliminated the bad effects of pesticides while we are continuing epidemics among the pests.

The other method of controlling pests is to raise insect allies that cause distraction but which in themselves are not harmful. The use of the Vedalia beetle or ladybug to control citrus cottony cushion scale has been mentioned and is a good example. Three tiny species of wasps are being raised at present to destroy the cereal leaf beetle. This parasite was introduced into the United States in the mid-1950's and is a serious threat to wheat and the other small grains. Of the wasps being used, one attacks the eggs of the beetle and the other two attack the larvae. Other examples exist, but will not be discussed.

Herbicides are the last of the biocides we should examine. These are compounds that kill weeds and some are quite dangerous to the human. During the Vietnam War, much of the fighting took place in the jungle. The military believed that if they could defoliate the jungle areas it would be much easier for our attack helicopters and personnel to see the Vietcong troops. Agents used included 2,4-D and 2,4,5-T, which affect broad-leafed plants; cacodylic acid for the destruction of elephant grass and rice; and picloram. The spraying was started in 1961, and by 1967, 150,000 acres of cropland had been sprayed and 500,000 acres of jungle defoliated.

There are several difficulties associated with the above agents. 2,4,5-T and certain of the others can probably cause tumors, miscarriages, and birth defects. Dow Chemical Company, which was the largest manufacturer of the agents used in Vietnam, was attempting in March 1979, to consolidate all of the individual suits into Federal Court in Garden City, New York. Up to April 11, 1979, 450 veterans have filed claims for injury sustained in Vietnam due to the herbicide

Agent Orange, which contained amongst other things dioxin, which is a highly toxic compound found in 2,4,5-T.

There is at present a great argument as to whether 2,4,5-T and Silvex are safe herbicides. Both of these herbicides contain dioxin. Not only can dioxin kill, but in even miniscule amounts it can produce cancer and birth defects; in laboratory animals, in minute parts per trillion, it also causes spontaneous abortions, testicular atrophy, and aplastic anemias. Some judges and agencies have permitted the use of one or the other—others have barred them.

In June 1977, Oregon protesters against the spraying of the timber-land near their homes went into a reforested area owned by Publishers Paper Company (a subsidiary of Times-Mirror Company) to prevent helicopters from spraying. This was successful. However, defying the warnings, the protesters again went into a brushy mountain site near Rose Lodge. Twenty protesters were arrested, and the others were sprayed with 2,4-D. Three Publishers Paper officials acknowledged that the demonstrators had been sprayed and that orders had been given to spray the area even if demonstrators were present.

The women of Oregon got both 2,4,5-T and Silvex barred in that state on March 10, 1979, due to possibility of miscarriages; since 130 per 1000 had occurred in Alsea, the test area, and only 46 per 1000 in the control areas. California advised against the use of 2,4,5-T on June 7, 1979, because it contains dioxin, and Mendocino County of that state banned its use on the same date.

Federal investigators have found that a ninety-three acre landfill site near Little Rock, Arkansas, was used for a deadly pesticide waste formed in producing 2,4,5-T for thirty years. This material was dumped by the Vertac Corporation in a landfill at Jacksonville, according to the Los Angeles *Times* of May 21, 1971.

A mile-long stretch in Monroe, Louisiana, was contaminated when a crop duster hired by the city sprayed this stretch with paraquat, an herbicide used to wipe out the marijuana crop. Department officials thought the damage was only slight to moderate, although there were long stretches of dead and dying trees and gardens.

In July 1976, an explosion at the Icmasa chemical plant in Seveso, Italy, blanketed the area with dioxin. Dioxin being a known deadly poison, the contaminated area was treated like a leper colony in an advanced stage of disintegration. Prior to a formal order issued to evacuate the area ten days after the explosion, homes full of furniture were burned, and 70,000 animals were destroyed. Within the first year, deaths from sclerosis of the liver doubled; no one can predict how many deaths from kidney and liver damage, leukemia, and cancer will result in the future. As of this moment, a six-foot wall encloses Zone A; it is an area of death. Only workers wearing special clothing, gloves, and masks may enter that zone; the 288 persons who formerly lived there may not return to their homes or gardens. Although topsoil was stripped from the area, dioxin continues to permeate everything which was exposed to it.

In the state of Michigan, no explosion—but an equally culpable carelessness—resulted in the poisoning of the motor state. The Michigan Chemical Company manufactured, in the early 1970's, a highly toxic fire retardant called Firemaster. It was made with PBB (Polybrominated biphenyl), a known carcinogen. Persons working with it wore special clothing. In 1973, Michigan Chemical also manufactured a feed additive called Nutrimaster. And, in that year, a vast quantity of Firemaster was shipped out as Nutrimaster.*

This poisonous material was sold to farmers throughout Michigan and fed to cattle. These animals developed abscesses, their hair fell out, and many became blind. Calves died and were autopsied, disclosing the toxic PBB. But many animals already had been slaughtered and their meat sold throughout the state; consumed by Michigan residents, the poisoning showed up in the form of dizziness, stress, and ulcerations. Other carcasses were rendered into meat meal and fed to animals, which were thereby poisoned; manure from infected cattle has permanently contaminated the topsoil of feedlots and pastures, continuing a permanent cycle of toxicity. Millions of persons have been, or may in the future be, poisoned as a result of this single episode.

In August 1981, the Michigan State Health Service announced that most of the fish from Lake Ontario for the previous three years contained unacceptable amounts of dioxin.

In 1972 the federal government ordered that all children's sleepwear be made flame-retardant. Two compounds were produced. The most popular was a compound named Tris, but this was barred when the compound was found to be carcinogenic in animals. A substitute, Fyrol FR2, was then used but this compound may also produce cancer.

Cattle have been killed by chewing on fence posts impregnated with pentachlorophenol used to preserve the posts. One can begin to wonder—is anything chemical truly noncarcinogenic?

*This is well discussed in the article by Rakstis, "The Poisoning of Michigan," *Reader's Digest*, September 1979, pp. 104–8.

Many other biocides are dangerous to health and life. The military would like to ship certain of their nerve gas bombs that are leaking, presently in Colorado, to a different reservation in Utah. These have now (August 1981) been sent to an isolated military camp in Utah.

At the present time, biological agents are being sought against weeds, as they have been against insect pests. Geese have been relatively effective in controlling nut grass and other weeds in cotton fields.

In California, a beetle has been fairly effective in the control of Klamath weed and another has been used against native sagebrush. Australia has been controlling cactus with the cactus moth, and many other examples exist. This should be the method of the future, in reducing dependence on dangerous herbicides.

There is one suggested biological agent about which I retain a healthy skepticism. The U.S. Navy plans to kill 4000 goats on San Clemente Island, as it claims they are destroying the brush and cover used by the loggerhead shrike and other species nearing extinction. Two organizations, The Fund for Animals, Inc., and GOAT (Give Our Animals Time), would like the goats removed to the mountainous brushy areas of Los Angeles County. They believe these would act as biological agents to keep down brush, helping to prevent fires.

Since we have seen the goats on San Clemente increase from 1500 at one time to 16,000, the same could happen in our foothills. This would mean that the brush holding the watersheds would be destroyed and erosion would be rampant. Another major, but not insurmountable, problem would be getting the goats to the mountain. Perhaps the proponents could use telekinetics, or call on supranormal forces. As Sappho said:

> Evening star, go bring all things which the bright dawn has scattered:
> you bring the sheep, the goat, you bring the child back to its mother.

The Soils We Till

Nations are wealthy or poor because of the lands they own. If the soil is rich, the people are well fed and content. Except during periods when temporary wealth may be obtained from the sale of an exhaustible resource such as oil, if the soil is poor, either through the activities of nature or of man, the people may beg for bread or starve in the streets. Women may become prostitutes if others will take them into their households and feed them. Men may use the knife or the strangling wire to obtain sufficient pennies to purchase a loaf of bread.

We will not speculate on how the earth was first formed but assume the inside is a viscous, hot liquid of which the upper surfaces have cooled sufficiently to become great plates upon which the continents, islands, and oceans rest. Actually, they do not rest, as there are stresses and strains as these tectonic plates slide toward and against each other.

There is movement, always movement, and slowly but surely, a millimeter at a time, California is moving towards Hawaii. The restlessness results in great land faults and earthquakes.

There is a theory that when the earth was cast off to become the third planet from the sun, it was a molten ball and all of the water existed as vapor far away but within the pull of earth's gravity. As the earth began to cool on the outside, a shell was formed that was changed many times over the eons. Mountains were thrust upwards only to disappear as new ones were formed. Finally, the earth was sufficiently cool so that one great land mass was formed. According to the believers in this theory, the land mass drifted apart into the present continents, and it is possible to see how these might have fitted together if one uses a little imagination.

Temperatures had cooled so water was no longer merely a vapor phase but fell to the earth filling the great voids that became the oceans, seas, and major lakes. Many changes occurred with the tilting of the earth to the north and the south. The angles were not always the same, and it is quite easy to demonstrate that the northern magnetic pole was not always where it is now, but that it changed location at least several times.

The rock that was thrust up to the surface differed greatly from place to place. That is one of the several reasons our soils are so different now, as some were formed from parent material rich in plant nutrients while other soils were formed from rock or parent material poor in plant nutrients. Swinburne expressed his ideas beautifully:

> Before ever land was,
> Before ever the sea,
> Or soft hair of the grass,
> Or fair limbs of the tree,
> Or the flesh-colored fruit of my branches,
> I was, and thy soul was in me.

The wealth of a nation lies in its soils and the deposits that are under the soil. Six hundred million years ago, single-celled plants and animals and then multicellular plants and animals formed the deep oil deposits of the present time. This was back in the Cambrian and pre-Cambrian eras. In most cases the tremendous pressure needed to change the plant or animal tissue into oil was derived from sedimentary rock which was formed and deposited upon these materials. In certain cases folding of the rock or deposits of lava dust blown on top further increased the pressure. As the eras changed, this was repeated on occasion and is the reason that oil deposits may be found at two or three different levels separated by some thousands of feet.

We could turn the clock back 280 million years ago to the Pennsylvanian period of the Paleozoic era when there was lush vegetation during this whole age. Temperatures were warm and the mean temperature of the area from 40 degrees to 90 degrees north latitude averaged 50 degrees F, instead of the mean temperature of 36 degrees of this zone at the present time. During these early periods, trillions and trillions of tons of plant material were formed and covered, and became the great coal layers and certain of the oil layers which were formed under vast pressure.

At one time, there were no ice-covered north and south magnetic poles. It is relatively easy to determine the temperatures that existed in different areas by the temperature sensitivity of the plants and animals found in these various areas. Thus, areas that are now our Antarctics and Spitsbergen are underlain with deposits of coal which indicate that these were once warm. Plants such as peppers and breadfruit grew up into what is now the far north, and animals that are linked to tropical areas such as alligators, crocodiles, and tapirs are found as fossilized remains in Alaska.

Let us now skip some hundreds of millions of years so that we can enter the age of glaciations which covered Canada, parts of the United States, and Europe, and gave us the great soil types found in the land or soils we till.

The earliest glaciation we will refer to is the Nebraskan, which started a million and a half years ago and lasted for a hundred thousand years. Exactly why glaciers form no one knows, or perhaps, many people know but their ideas are completely different. It is of no use for us to discuss the four major themes which differ completely, for if one is correct it means that the other three are, of necessity, incorrect. I personally believe that there was a change in the magnetic poles that caused the first glaciers to form. In the deep mountain valleys the sun did not penetrate, and these were cool or cold all year long. The mean temperature dropped, due to the angle of reflection of the sun. Snow and ice accumulated in the deep mountain valleys, for old snow becomes ice. Glaciers tend to form their own climate since the brilliant white snow reflects more of the heat of the sun from the earth. Once the ice sheet grew to be over a thousand square kilometers, a new climate of its own was formed.

The great glaciers cooled the air and inched farther and farther south. They moved slowly, but relentlessly, over the soil in much the same way as an anaconda moves upon a hypnotized deer. Finally, this first great glaciation included twenty million square kilometers of land area and probably ten million square kilometers of rock ice.

As the great ice sheet moved southward it acted as a great plane, leveling the hills, scooping out great basins that could become lakes, and grinding the rocks into a flowing mass intermingling these together. The forward movement of the glacier stopped when the inrushing warm air from the south melted the outer face and pushed the ice backwards. The farthest advance of rock and pebble by the glacier is called the terminal moraine. During the hundred thousand years the first glacier that we call the Nebraskan was moving south, rain was falling upon the unglaciated portions south of the glacier and dissolving food nutrients from that soil, leaving it poorer than the glaciated area.

There was a period of 200,000 years in which the glaciers existed only farther north in Canada, and the land of Wisconsin through southern Illinois was all exposed to the leaching impoverishment of rainfall. Then the Kansan glaciation moved down. Not so far and not quite in the exact areas, but it plowed new rock from below the surface of what had become soil over the 200,000 years, and redeposited a new layer of glacial till enriching the soils farther to the north.

After a long period of time there were two more glaciations—the Illinoisan and the Wisconsin. The latter disappeared only ten to fifteen thousand years ago, and, as it reground new rock, it left the very fertile soils of central Illinois, Wisconsin, and certain of the adjacent states. It finished scooping out the Great Lakes. The soils so recently uncovered by the Wisconsin glaciation are some of the richest soils in the world. The soils are teeming with both microscopic and macroscopic forms of life. Man has not had time to impoverish them, whereas to the south the land has been exposed to the action of the elements for hundreds of thousands to millions of years, leaving soils that are poor in nutritional capability. As Auden would say:

> The glacier knocks in the cupboard,
> The desert sighs in the bed
> And the crack in the tea cup opens
> A lane to the land of the dead.

In classifying the lands we till, it is best to start with the oldest. These are the soils that were unglaciated, warm, and exposed to high rainfall of approximately eighty inches; and these soils lie in an area 23 degrees north to 23 degrees south of the equator. They are called laterites or soils of the tropical rain forests. The vegetation appears lush, but the rainfall has leached out almost all of the potassium, calcium, and magnesium nutrients. Even the silica has been solubilized so that, except for a few inches at the top where limbs and leaves decay and these minerals are recycled, there are only iron oxides and aluminum oxides. The Khmer civilization of Angkor Thom had walls, palaces, and temples of laterite. The peasants were able to grow crops upon this soil until the cities grew so large the organic matter was destroyed, and the cities fell before the ravages of nature and barbarian tribes.

The Mayans of Guatemala and Honduras fell because the wise men who could calculate a perfect calendar and evolve a system of mathematics could not see that there was little soil on the surface of the laterite. The Mayans had succeeded other ancient but cultured tribes who had not farmed intensively so therefore did not wear out the soil. As the great Mayan city-states grew, the farmers had to move farther and farther away as the black topsoil was oxidized. Finally, it became necessary to migrate north or starve. A few years ago, at Tikal, the topsoil again looked black, as the jungle had returned. Yet, with my shoe, I could scrape through the debris and expose the laterite.

Within the last three generations, many people thought the Amazon Basin of Brazil would be the breadbasket of the world. They did not see the brick red laterite exposed through the black topsoil. The city of Iata was built. The trees of the jungle were bulldozed and burned, destroying almost all of the organic matter. Wheat could not be produced after the first few years. The former city is now a small river village. According to the Los Angeles *Times* of May 7, 8, and 9, 1979, the last of the rain forests of Brazil are disappearing as man destroys the cover for agriculture. May the land rest in peace!

In northeast Brazil the floods sweep over the laterite, wiping out villages and forcing the peasants and the Indian aborigines to flee south, looting the stores of the cities to obtain food to replace their destroyed cattle and crops. Of the thirty million people of the area, eleven million are severely affected by the floods and droughts.

A similar situation exists in the Mekong River Delta of Vietnam. For centuries, the river has been carrying organic alluvial material, depositing it in the delta area, which is laterite, and making it the rice bowl of the Orient. If the Mekong River Dam is completed, that will stop the flow of this silt material, and, in time, the nonfertile laterite soil will be exposed.

A close correlation exists across the equatorial belt of Africa. The United States has sent assistance in the form of persons who do not understand laterite. In Ghana, the Volta Dam completed in 1969 not only has ruined the cropping system, but also has caused river blindness in natives. In the Ivory Coast, the Kasson Dam has done the same, and the Kariba Dam in Rhodesia is interfering with agriculture as well as displacing many natives and increasing the spread of disease.

There are few ways of handling laterite soils. The primitive cut-and-slash method for small populations in which the villagers grow yams and plantains for a few years, then move on and permit the jungle to recover the area, is one. Another is to put the land in mixed grasses and legumes and use it for pasture or hay. Or the land can be put in bush crops such as the coffee bean. In these cases, the land should be covered at all times. When this is not done, Hodgson would say:

> I saw in a vision the worm in the wheat,
> And in the shops nothing for people to eat;
> Nothing for sale in stupidity street.

A semirelated type of land is the soil that would be alkaline or saline in the absence of sufficient water. For thousands of years, the Nile River carried sufficient water so that the peasants could fill their irrigation ditches after they had allowed the water to overflow the land. Sufficient rich silt was carried in the waters to keep the lands fertile, and the excess silt with its rich nutrient salts flowed into the Mediterranean Sea where it furnished food for the great sardine and tuna fisheries. When the Great Aswan Dam was built to furnish power and control irrigation, the silt no longer flowed to the sea, and is now precipitating out in the back of the dam, filling it up. As a result of lack of food, and of pollution from industries and cities, the fisheries of the Mediterranean are almost destroyed.

An argument still rages in 1981–82 as to whether or not the Great Aswan Dam is worthwhile. At least 30,000 tons of Nile perch are caught annually in Lake Nasser, which balances the previous Mediterranean fish tonnage. Microscopic blood flukes have increased and 70 percent of all Egyptian males are infected with the debilitating disease known as bilharziasis or schistosomiasis. It might be mentioned that only seven of the twelve Russian-made turbines that supposedly would produce eight to ten billion kilowatts of electricity can operate simultaneously at the dam.

In the Soviet Union, which is the world's leading producer of cotton (most of which comes from the irrigated desert lands of Uzbekistan) the water used for irrigation is lowering the level of the Aral Sea, the fourth-largest inland body of water on earth. It, like Mono Lake of California, is drying as the fresh water needed to replenish it is diverted to irrigation. A complex canal system is planned not only to irrigate the cotton crop, but also to leach the saline soils, which in 1981 are stunting the plants and diminishing the crops.

A great area of West Pakistan is being converted to saline (salty) soils. The once mighty Indus River, which amazed Alexander the Great, has been diverted to canals for so long that there is not adequate water to wash the salt, being carried to the surface by capillary action, deep into the soil profile where it will not injure crops. This is somewhat similar to the desert soils of the United States, where almost no living things grow. Over the centuries there has been just sufficient rainfall to carry salt or calcium deep into the soil. If there is any underground moisture, similar salts rise and a layer of hardpan forms where they meet. These are called solonetz or solonchak soils.

In the Arctic where the land is frozen, the tundra exists. The subsoil is permanently frozen, and, at best in the short growing period, berries and other plants with a short life cycle can be grown.

The extreme northeastern portion of the United States has true podsol soils. There has been leaching of the fine clay and nutrients from the surface into the layer below, known as the B horizon, which is a cemented hardpan. The soil is acid and was covered with coniferous trees, such as spruce and pine. When these were cut, only a few crops would grow. Potato and small dairy farming are practiced in the Aroostook Valley area of Maine.

The Pilgrims landed somewhat farther south where the soil was podzolic. It was acid, stony, and not very fertile. The stones were used to build stone "dykes" or walls around the fields. The farming conditions were poor but tolerable, and when the early pioneers heard of the prairie soils of Ohio, Indiana, and Illinois, they were willing to subdue the lands and the Indians through the plow and the Kentucky long rifle. These soils, particularly of northwestern Illinois and Iowa, were the best of those in the United States.*

As later groups of farmers moved farther west, they found the black earth or chernozem soils. Like the steppes of Russia they were rich, but

---

*As stated in *Reader's Digest*, August 1979: "Man, despite his artistic pretensions, his sophistication and many accomplishments, owes the fact of his existence to a six-inch layer of topsoil and the fact that it rains."

the rainfall was not sufficient to produce good crops, so they moved farther and farther west. The land was excellent and rich in nutritious buffalo grass, but when it was plowed and put into wheat, moisture was lacking. Possibly two crops of wheat could be obtained; then the land would have to lie fallow (bare) for a third year to collect more moisture.

The soils of southern Illinois, and other areas of southern United States, had been exposed for hundreds of thousands of years to rainfall and were lacking in mineral nutrients and nitrogen. Carbon, derived from decomposing trees, green plants, and other vegetation, is of equal importance with nitrogen. Where the true prairie soils needed only light liming and fertilization, the soils farther south needed heavy liming, fertilization with most of the nutrients, and the growth of legumes so that the bacteria associated with them could take nitrogen from the air and return it to the plants and soils.

The soils of the United States were poorly handled and suffered greatly from erosion. The fertility had originally been based upon the parent material (type of rock) converted to gravel, then sand to silt, and eventually clay, which has the finest particulates.

Of approximately 1900 million acres in the contiguous forty-eight states, 702 million have not as yet been badly damaged by erosion, 775 million are moderately damaged, 145 million are not determinable as mountain and desert, and 282 million acres have been completely destroyed. Part of the above soils are washed by rainfall to delta areas of rivers where they become structureless; that is, there are no horizons or layers as a new deposit is formed each year. As Goldsmith said about similar ruins:

> Ill fares the land, to hastening ills a prey,
> Where wealth accumulates, and men decay,
> Princes and lords may flourish, or may fade,
> A breath can make them, as a breath has made;
> But a bold peasantry, their countries' pride,
> When once destroyed, can never be supplied.

Nor can the basic soils on which a nation must depend.

William Byrd, a gentleman farmer of the Piedmont area of Virginia, noted the terrific erosion around the corn and tobacco plants on his farm. Patrick Henry believed that some of the greatest patriots were those who controlled erosion of the soil. In Stewart County, Georgia, once noted for its good farmland, one-fourth has eroded away. About three-fourths of the state of Oklahoma has been somewhat eroded, with six million acres suffering quite badly. Indiana has had severe erosion of 40 percent of its soil and much erosion has occurred throughout the Midwest. In the Pacific Northwest, which is a young agricultural area, 35 percent of the land is subject to severe soil blowing. In this area known as the Palouse Region an inch of soil is lost to erosion every fifteen years due to the carelessness of man. It took nature 700 to 900 years to form this amount of topsoil.

Our best soils are being lost to water and wind erosion, to urban development where houses cover the former cropland, and to the concrete jungles we call expressways or freeways. Annually, we lose six billion tons of topsoil per year from erosion and our covering the area of several large cities with homes and concrete.

In order to prevent erosion, the modern farmer should plant his crops on a contour, rather than up and down the slope. He should use strip cropping, if necessary, which would be wide bands of grass or legumes between the wide bands of corn and soybeans to stop erosion. Every farmer should follow a good crop rotation system. This means that he should decide upon a three- or four-year rotation in which crops are grown in sequence upon the land, on a contour if the land is hilly. One section might be in corn one year, followed by wheat or barley in which clover or alfalfa is seeded. The third year this would be used for hay and pasture after the wheat is harvested, and then returned to corn, cotton, or soybeans, depending upon where he is farming.

If the lands are steep, they should be terraced, as are the vineyards of the Rhine Valley, the Burgundy region of France, or the rice fields of the Orient. Cover crops should be used wherever possible so that the bare land is never exposed. This can be done by surface mulching, which is easily practiced by cutting the corn stalks at ground level after harvest and permitting them to lie upon the soil.

In north central Nebraska, a sand area of 20,000 square miles exists. It is a land of valleys, basins, and rolling dunes covered by sagebrush, yucca, and grasses. Other sand areas along lakes are developing in the United States.

In some areas of the world erosion was fought. Farmers in Antioch, Syria, built terrraces to retain their soil. In Ireland, stone hedges or terraces date back to approximately 2000 to 3400 B.C. In Peru, the Incas built stone terraces to retain their soil.

In 2000 B.C., northern Africa was an area of olive groves and wheat fields, and was plentifully supplied with food and water. Great cities had been built in Africa in 1000 B.C. in areas that are now desert. The Sahara Desert is marching relentlessly at the rate of a mile a year in many areas. The Great Thar Desert of India has increased 60,000 square miles in the last hundred years.

The difficulty is that man has been careless and contributed to erosion of the soil. In 2000 B.C., the Babylonians harvested two food crops a year. The whole area which was the Fertile Crescent, and the lands around it, are now largely desert. An area that was noted for fertility under Darius I, in Persia, approximately 2400 years ago is now mostly desert.

Less than 20 percent of modern Iraq is now cultivated. The old canals that irrigated the land are filled. The ancient seaport of Ur now lies approximately 150 miles inland and is buried thirty-five feet deep. The carelessness of man permitted nature to erode the soils, soils that at one time had great fertility. Iraq was not an under-developed nation; it was an overdeveloped nation that grew too soon, and lacked the wisdom to protect its soils against erosion.

The forests and fields of Homer are now badly eroded. Pliny discussed the beautiful hills of Greece and of Sicily. These same lands are now virtually barren, as they were overgrazed and eroded. Sheep and goats pulled up the grasses by the roots. When these were gone, they grazed upon the brush and girdled the trunks of the young, growing trees, destroying them and leaving the land unprotected. As Tennyson said:

> He clasps the crag with crooked hands,
> Close to the sun in lonely lands,
> Ring'd with the azure world he stands.
> The wrinkled sea beneath him crawls,
> He watches from his mountain walls,
> And like a thunderbolt he falls.

The soil is a thin skin surrounding the earth. It would be pro-portionate to the normal skin of an apple surrounding an apple 200 feet in diameter. In the last 2000 years, the area which must feed mankind has greatly shrunken. In just this century, our massive public works—designed to chain the rampages of rivers—have forced, and will in the future force, millions of people to lose their food resources. As populations continue to explode in numbers, the area available to feed our masses will become increasingly smaller. Treat those soils kindly; without them, we perish.

The Energy We Use

It is the year 2376. Eighty-one years ago the 455 trillion barrels of oil reserves known in 1982, and the 189 trillion barrels discovered later, are gone from the earth. Every well has run dry. The vast reserves of Kuwait, Iraq, Libya, Saudi Arabia, and those of the continental coasts have expired. The peoples of the Middle East, that region almost devoid of water, who had preyed upon the other nations with exorbitant pricing of their produce, disappeared first as they could not raise foodstuffs and none welcomed them as immigrants.

The coal reserves, which had been known to be 5000 million tons in 1982, and had been increased by new discoveries to be 8000 million tons, had been used primarily by gasification and liquefaction. The gas reserves had been expended much earlier; and man was already struggling with his diminished universe by 2281 to produce foodstuffs for the areas remaining. The "greenhouse effect" had produced the rising waters which had resulted from the melting of the polar ice caps and glaciers. These covered the areas that had been the low-lying oil producing nations, much of the fertile areas of coastal Europe, the low-lying areas of China, and other Oriental countries. In the United States, there was no Mississippi River Valley. It had become an inland sea; and the coastal states except for the portions 150 feet or more in height had disappeared as had the similar areas of Mexico, Central and South America, and Africa.

Let us examine the history of mankind as it existed approximately 400 years ago—as if we were living then, in 1982. The United States, with 6 percent of the world population, uses 35 percent of the fuel supply of the world, and Russia uses approximately half of that amount. Of the energy we use, 35 percent is for industrial uses, 25 percent for commerce and agriculture, 20 percent for transportation, of which half is private, and 20 percent for residential heating and other purposes.

Of the energy used in the United States,* petroleum furnishes about 43 percent, natural gas 26 percent, coal 19 percent, hydroelectric power 3 percent, and nuclear power 9 percent.** We have reserves of over 1500 billion barrels of kerogen if we could liberate it from the shale, and also some tar sand reserves. The United States contains sufficient coal reserves for two to three hundred years, but obviously we need a new energy policy to develop collateral sources of energy, including wind power, sea power, geothermal, and solar power, in addition to the fossil fuels. We will discuss nuclear power in a separate chapter.

This has become more important since we are seeing the attempted rape of the world by the OPEC nations on one hand, and an extremely confused executive policy and energy-czar administrative policy, on the other. Although we will have to depend to a great extent upon oil and liquefied natural gas for the next decade, by that time, it is hoped, we will see gasified and liquefied coal, the distillation of kerogen into petroleum products, and an intelligent management of our offshore and Alaskan oil resources rather than leaving these blocked by a confused and inept government.

An examination of wind power indicates that a small amount of power could be obtained in this way, but the windmills or wind turbines, of necessity, would be in areas where there are relatively strong prevailing winds. Most frequently, these are in mountain passes fairly remote from large cities. A small generator was installed in Vermont

*U.S. Department of Energy figures for 1977.
**The figures for coal and nuclear power have since increased, oil has decreased.

in the 1940's and a somewhat larger one in Denmark. NASA predicts that eventually this source could supply 5 to 10 percent of our power needs if sufficient storage were available. A New York firm had mounted the first pair of 200-foot blades on the world's largest electricity-generating windmill at Boone, North Carolina, as of April 1979. The windmill, which was a $3.5 million federal research project, is expected to produce 2000 kilowatts of electricity, which would power approximately 300 homes. Another firm will build a series of twenty windmills in Pachero Pass near the San Luis Reservoir in California. The contract was let in April 1979. In February 1979, according to one report, a Barry Lebost had produced a more efficient wind turbine than the windmill. Other machines have been used but are not very efficient.

Pacific Gas and Electric Company, in April 1981, announced it will install 146 wind generators in Sonoma County, which will produce 250,000 kilowatts of electricity. The Japanese are using a combination of wind and oil to drive the tanker Shin Aitoka Maru, at a saving of 40 to 50 percent of fuel.

Water power can furnish Norway with the electricity it needs. In addition to the waterfalls, were they ever needed, dams could be built across some of the fiords, which would permit the incoming tide to race inside when the great gates were opened. As the tide fell, the water would flow out through turbines producing power. The French government built a similar plant on the Rance River, but it was expensive and difficult to manage because a computer had to be used continually to change the pitch of the hydroturbine blades. A similar plant has been suggested for the Bay of Fundy but was not built because of these difficulties, even though a water height differential of six meters is established by tides.

A small ocean power plant producing fifty kilowatts was successfully operated from a barge off the island of Hawaii by Lockheed Corporation. In April 1980, the Senate approved a measure to double the $40 million currently spent on energy research. The emphasis would be on Hawaii and Puerto Rico, which at present must generate most of their energy by burning oil. A similar project using "pond power" was successfully demonstrated at Ein Bokek in Israel.

Another type of sea power generator has been suggested in which the differential temperature between the surface warm water and the deep cold waters would produce electrical current. One method suggested was to have a massive floating turbine generator turned by refrigerant gas heated and vaporized by the warm surface water. Below this could be large condensers where cold water drawn from 600 to 900 meters would condense the refrigerant to liquid so it could again vaporize and turn the turbines.

One plan being examined, in January 1979, by the Department of Energy was to tow a massive turbine approximately 200 yards long into the Gulf Stream on the eastern coast where the slow moving current of about four knots would turn the massive turbine blades. It was thought that 250 of these in the Gulf Stream off the coast of Miami could power much of Florida. The idea seems impractical based upon weight versus power and probably corrosion. If it did work, the flow of the Gulf Stream might be changed.

Geothermal power is the use of heat localized in the earth. Areas that could serve as geothermal reservoirs are limited but have possibilities where they exist. There are two factors that must be realized to understand the possibilities of this source of power. First, in many areas salt corrosion of metal piping would be high and blockages of thermal-resistant plastic might be frequent; second, the steam is usually low pressure rather than the high pressure of the nuclear reactor or fossil fuel plant, and the thermal efficiency is only a third of either of the latter. The water and accompanying gases, which frequently contain ammonia and hydrogen sulfide, need to be removed rather than released into the air. Geothermal sources were thought to be rapidly depleted and the efficiency decreased approximately 50 percent each five to six years. The Imperial Valley of California, the geysers' geothermal steam field in Sonoma County, California, and the area near Yellowstone National Park, are possibilities. It is thought that one million hot spots might exist in the United States, particularly in the five western states, which could furnish power equivalent to twenty-five nuclear power plants. The plants would be very expensive, much more so than the heating I saw in Iceland in which the geysers and hot springs are used. By the year 2000, geothermal energy might add sixteen Quads to the 110 Quads needed at this time.*

A third system of the use of geothermal energy is to use the heat in hot, dry, rock areas such as those at Marysville, Montana; the Valles

---

*A Quad is one quadrillion British Thermal Units.

Caldera, New Mexico; and the Crater of the Moon in Idaho. It would be an expensive proposition to develop the hundreds or thousands of dry wells necessary to furnish the heat, as water is not present to furnish a heat-transport system.

Solar energy is an inexhaustible resource in the lifetimes of hundreds of generations to come, should mankind solve its remaining survival problems. Less than 0.1 percent of the energy striking the surface of the earth is absorbed; the remainder is reflected primarily as long-wave radiation. A portion of the solar energy was trapped for hundreds of millions of years by green plants, part of which were eaten by animals, and, as long ago as the Cambrian and possibly pre-Cambrian eras, formed the bases of our oil deposits. Later trapping of sunlight formed the deposits of coal.

We can use wastes indirectly by separating all of our present combustible rubbish whether it be boards, paper, brush, grass trimmings, or garbage, and burning these (a process called biomass), utilizing some pulverized coal or fuel oil and producing vast amounts of energy. Any medium-sized city could produce all of its power for light, home heating, and air conditioning by burning its wastes as suggested previously. It might not be able to produce all of the industrial power needed and none of the gasoline, but it would save almost half of the power we use and need in the United States. It is better to take active steps than passively await a day of reckoning, somewhat as Burroughs describes:

> Serene, I fold my hands and wait,
> Nor care for wind, nor tide, nor sea;
> I rave no more 'gainst time or fate
> For lo! my own shall come to me.

A tremendous number of ways of using solar power directly, rather than indirectly, are available, and should be attempted at the present time. All new homes can be built with thermal collectors for sunlight. These are somewhat expensive, but most will pay off in that no power, except for a small pump, is needed to furnish all hot water for heating the home and for other purposes. The solar collectors are relatively inexpensive in areas such as the Southwest—more expensive in the Northeast. Basically, the idea is that in the hours of sunlight, a fluid is heated, usually water or oil. Concentrators in northern climates focus the sunlight on the collector. When the temperature drops, the fluid is passed into the heavily insulated thermal storage area. If water has been used, it can be circulated directly through radiators or used as the hot water source. When the sun again comes up in the morning, part of the storage water and new water is shunted into the heater area by a thermoswitch. In Maine and other areas it might be necessary to assist in water heating with propane.

It is rather surprising, however, that with minimal sunlight focused mirrors can increase heating almost beyond belief. Solar heating is a relief measure to remove part of the load from power plants to homes and even to offices. Other and more efficient solar energy approaches are available, if man uses his ingenuity; and this is an area which should receive much more encouragement, and financial aid, from the government.

Solar batteries located on earth or even on a satellite probably will not furnish energy for electrical power for industry. The cost of solar batteries is tremendous, and, if these were located on a large satellite, the expense would be enormous.

The immediate future of the United States lies in its ability to produce "liquid coal" and to develop its own oil resources. We use approximately two to three thousand calories of heat per capita per day for our bodies as food, but we need 300 thousand calories per capita per day for our industries and our cars. It could be interesting to note how the calories per day increased from the time of early man, until he learned to use fire for heating, up through the industrial revolution, to the vast energy required today.

If we were forced to do so we could, by so designing our power plants, have sufficient coal for the next 350 to 500 years at our present rate of usage. This would mean, were there no oil sources, we would depend on electrical sources of transportation such as trams, trains, and even electric automobiles for short runs. The future would not be as black as the coal itself if coal liquefaction and gasification were used. These programs are already in research and pilot plant stages, and we could produce our fair share of the world production of forty trillion kilowatt hours per day from coal and its by-products if needed. As we will see, however, we do have oil resources* but have to combat a shortsighted government to be able to use these. As Burke stated:

*In view of the importance of petroleum in the manufacture of hundreds of products, including petro-chemicals, plastics, nylons, etc., it seems important to find alternate energy sources instead of simply burning up these reserves as fuels.

> Public life is a situation of power and energy; he trespasses against
> his duty who sleeps upon his watch, as well as he that goes over to
> the enemy.

Our energy crisis is real, and the coal-fired 3000-megawatt plant in
Nevada, and a similar plant in Utah, ought to be built. In California, the
construction of a 1600-megawatt coal-fired power plant in the Sacramento Valley, which was proposed by the Pacific Gas and Electric
Company, is being blocked. There have been probably at least fifty
coal-fired power plants opposed by the United States Government or
state bodies.

Environmental groups are bitterly opposing the construction of
coal-fired power plants and oil refineries, and this is one of the reasons
we have a power shortage. It doesn't seem to make much difference if
these plants were to be far away from cities, or close to the coastal
areas—a different reason has been found in each case. I am a conservationist and an environmentalist, but I am not a fanatic. Most of
the persons, actually all of the persons I know who fight these bills,
drive to the meetings instead of walking or riding a bicycle. In January
of 1979, the EPA wanted Pacific Gas and Electric Company to build a
coal-fired power plant near Travis Air Force Base, rather than close to
Collinsville near the Saisun Marsh. This would not only have placed
two 600-foot high smokestacks under a flight pattern from the air base,
but would have removed it from a source of cooling water. This is
typical of bureaucratic action; fifty references of like myopia could be
given.

The modern coal-fired plant with efficient scrubbers in its stacks
can be as clean as any factory. Liquefaction of coal seems, however, to
be the most reasonable way to escape the present energy crunch. It
would be a secondary step that could supplement our oil supplies, with
probably the government assisting rather than damaging the program,
as it has with the oil situation as it exists in this country. Liquefaction
of coal is complex but it, and gasification of coal, is now in pilot plant
stages. The end products are different from those of petroleum. With
coal we would secure approximately 50 percent fuel oil, 30 percent
turbine fuel, and 15 to 20 percent gasoline; whereas with petroleum we
can get almost 60 percent gasoline, 30 percent turbine fuel, and 10
percent fuel oil. If we can convert to a great coal industry in the next
decade, there would be no dependence upon the OPEC nations.*

Liquefied natural gas (LNG) is an excellent source of energy. In December 1978, however, the Energy Department denied El Paso Eastern Company and related associates a proposal to import LNG from Algeria by way of Texas. The same agency denied a proposal by Tenneco to bring such gas from the Bay of Fundy. Both Hazel L. Rollins, in the first instance, and David J. Bardin, in the second, assumed our natural gas supplies of the United States to be adequate, it was reported. There seems to be no understanding on the part of the government that supplies in the United States may be inadequate, when looking into the future.

There was one accident with liquified natural gas at Cove Point, Maryland, in October of 1970, but the House Committee has approved the alteration of plans that would make these plants safe.

Gasohol, the mixture of approximately 10 percent alcohol and 90 percent gasoline, has been used as a substitute for gasoline. Either methyl alcohol or ethyl alcohol can be used. Methanol can attack the lining of gas tanks, swell the filters, and clog engine components. Ethanol is not as active in this manner. Methanol has only about half of the energy value of gasoline whereas ethanol has two-thirds of this value. If water is present, even at 0.1 percent to 0.2 percent, the mixture can separate into two phases, and efficiency is minimal compared to gasoline alone.

President Joao Baptista Figueirado of Brazil in 1980 uses an alcohol-powered automobile. That nation hopes to eliminate the need for gasoline by 1990.

The petroleum crisis of 1979 was minor compared to those crises one could expect to occur later. At that time, industry was growing in the world, particularly in the United States, and the producers of oil needed a leader as president, rather than a lecturer unaware of the problems.

Unfortunately, that administration increased the time lag for new oil discoveries. In Alaska, there are probably seven offshore frontier areas high in gas and oil, but relatively small amounts of these areas will be leased. There are blockages against the lease sales, such as those for the Georges Bank, originally scheduled for sale in January 1978, but delayed by a last-minute legal challenge. The outer continental shelf (OCS) is one of the most promising sources of oil and gas,

---

*In 1981 the United States has 786 billion tons of coal reserves and uses only 786 million tons of coal annually.

and still only 5 percent has been leased. This 5 percent accounted for over 15 percent of the oil, and 23 percent of the gas, produced in the United States in 1977.

In the United States there are 150,000 gas wells.* The Federal Power Commission (FPC) was created in 1938, supposedly to keep the prices "just and reasonable." The commission first tried setting prices on individual wells, but this was not possible. In 1960, it adopted wide-area pricing which was sustained by the Supreme Court in 1968. The commission kept the prices fixed at approximately the 1960 levels, and, as inflation grew, production became more costly. It became necessary for the producers to sell intrastate where FPC had no control. Gas prices then rose intrastate to three or four times the interstate price. In the 1970's, consumers tried to get natural gas at the 1960 price instead of petroleum at the OPEC price, but it was not available. The Energy Resources and Development Administration (ERDA) estimated that, at prices between two to three dollars per one thousand cubic feet, there was enough gas to last for a thousand years at the present rate of consumption. From newly developed fields, however, only approximately $2.70 can be obtained per thousand cubic feet.

With respect to crude oil, in 1973 all oil import quotas were ended, and, at about the same time, depletion allowances were ended for major oil companies. Both made domestic crude oil drilling and production less profitable, although the number of drilling rigs searching for oil in the United States climbed to 2777 in May 1980, an increase of 43 percent in one year. At the time of the Arab oil embargo, the refiners could not afford to produce gasoline under the artificially low controlled price; the oil went into other uses, or the plants were shut down. By mid-1974, the price controls of refined products were starting to be eliminated but those of domestic crude oil were maintained. Oil prices from OPEC increased from $3.00 a barrel in 1970 to over $35.00 in 1981.

In November 1973, the Emergency Petroleum Allocation Act (EPAA) provided for a two-tiered pricing system for domestically produced oil. This was "old oil," or that produced from domestic fields during the corresponding month of 1972, and new oil from fields developed after 1972. Producers were permitted to release a barrel of old oil from its classification for each barrel of new oil produced, beyond the base level of the old crude.

*We have reserves of approximately 5000 Quads of gas and use 56 Quads annually.

The maximum price of old oil was set at $5.03 per barrel, plus $1.35 subject to readjustment. Imported oil, new oil, and stripper oil (oils from wells producing less than ten barrels per day), were allowed to sell at market price. Regulations became more complex over the years and such gimmicks as the Entitlement Program were introduced to supposedly equalize profits disparity between refiners. Under this act, the importation of foreign crude oil is subsidized. It pays a refiner to purchase more foreign oil and increase his production of refined products. Energy subsidies, taxes, rules, and prohibitions are growing at a tremendous rate. James Schlesinger, former Secretary of Energy, stated in *Journal of Law and Economics*, October 1968:

> The tool of politics, which frequently becomes its objective, is to extract resources from the general tax payer with minimum offense, and to distribute the proceeds among innumerable claimants in such a way as to maximize the support at the polls. Politics, so far as mobilizing support is concerned, represents the art of calculated cheating—or more precisely, how to cheat without really being caught. Slogans and catch-phrases remain effective instruments of political gain. One needs a steady flow of attention-grabbing cues and it is of lesser moment whether the indicated castles in Spain ever materialize.

Windfall taxes occur in every industry. There are a number of industries, including cosmetics and drugs, that have greater percentage returns on investment than does the oil industry. Even the individual who purchased a home for $20,000 in 1955 and sold in 1981 for $80,000 has made a windfall of 400 percent. This the government permits him to keep as a once-in-a-lifetime negotiation. Former president Carter proposed to tax windfall moneys in the oil industry at 50 percent and Congress, with such individuals as Senators Udall, Hart, and Kennedy, seems to support a tax of 80 percent.

This seems the wrong way to handle the situation. The United States needs new explorations in Alaska and off the eastern and western coasts, and a large number of new refineries. What should be done is to leave these funds with the oil companies for these purposes, making certain that the funds are so utilized, or for placing older refineries into a state where they will meet all demands related to air pollution.

Even here, every state and federal agency should work together to assist the companies in the building of the new plants, not stalling them with one regulation after another, so that it takes years to obtain the proper permits. Also, all funds spent for the development of new fuels such as liberation of kerogen from shale, and its fractionation into heavy and light oils, as well as projects of liquefaction and gasification of coal, should be part of the legitimate use of "windfall" taxes.

Conservation of power must be practiced by all. This includes the purchase of less powerful cars when our present ones are worn out. It demands that sane driving practices be maintained with speed limits, as at present, of fifty-five miles per hour. Homes in winter should be maintained near 65 degrees F to 70 degrees F; and, in summer, air conditioners should not be used until the temperatures exceed 80 degrees F. Lights should not be burning in empty rooms but should be extinguished when they are not needed. We do not have sufficient power, and, unless we conserve it, catastrophic years move more quickly upon us. In 1981, incentives are being offered to the builders of energy-efficient homes and to those who convert their homes to be more energy efficient through insulation, double glazing of windows, storm doors and other devices.

One major point that is of great concern is the problem of oil spills. So many of these have occurred, and so much damage has been done to both plant and animal life in the ocean, that drastic steps should be taken.

I have in front of me, as I write, the record of at least a hundred oil contaminations or spills covering a number of years and a number of sources. Some of these are fresh water contamination, but the major portion are marine problems. The very earliest were natural contamination, or seepage from the sea floor; in 1793, Captain George Vancouver noted the slow release of gas bubbles and tars at Santa Barbara, California. Most of the spills have been the result of accidents, although a few land-based spills have been deliberate.

Since two-thirds of the petroleum used in the United States is carried by tanker, there is the possibility of accidents both near our coasts and in other portions of the world. Tanker sizes have increased from the T-2 of World War II with 16,000 dead weight tons, through averages of 27,000 tons in 1965, 76,000 tons in 1966, 119,000 tons in 1967, 312,000 tons in 1970 to the 540,000 ton super-tankers of the present time. Not only can oil tankers pollute the waters, but also, in collision of dry cargo ships, the bunker fuel oil can pollute the area.

As far back as 1966, the California Fish and Game Department reported 181 oil spills in the Los Angeles and Long Beach areas alone;

of these nineteen were land-based, fifty-nine were merchant vessels, and sixty-seven were naval vessels.

I will mention the types of land-based spills first, which include the accidental and deliberate. Most cars leak small amounts of oil which fall upon the road surfaces. In Los Angeles, when it rains, this oil is washed from the freeways into the main drains. For the first hour, the roads are unusually slick. This oil then usually passes through the sewage plant where some may be processed. The balance enters the ocean. The oil waste industries dump oils they cannot salvage into the sewage system, and the brine waters containing oil may seep into the ocean from the sloughs surrounding the wells.

Ruptured pipelines on land account for spills, and in May 1979, 400,000 gallons of diesel oil poured out at Big Durbin Creek in South Carolina, forcing the towns below to close the intake valves of their water plants, and, incidentally, resulting in large fish kills. It was determined, on May 16, that the total spill was of 750,000 gallons. In June 1966, a sewage plant in Colorado discharged 20,000 gallons of used oil. In the same year, a levee around an oil company holding pond broke, discharging 200 barrels of crude oil. Prior to this, in 1962, a pipeline broke, discharging 1,400,000 gallons of xylene and light mineral oil into a Minnesota river from a storage facility.

Once in a while justice prevails, and one newspaper noted that a Martinez court judge, in California, had sentenced an individual to fifteen days in jail for polluting a creek with waste from his two oil wells. This individual had been on probation because of a prior incident of discharging oil into the creek, which caught fire in 1976.

These are minor compared to some of the barge and tanker spills. The year of 1978, which witnessed a number of spills, ended badly when on November 1 a Greek tanker was wrecked off the coast of Wales. About 26,000 tons of oil were salvaged, but when the ship went down, between three and four thousand tons of oil had been spilled. Great damage was done by an oil spill of the tanker *Esso Bernica*, which spilled a thousand tons of oil in the Shetland Islands during the last part of 1978. Since that time, it has been discovered that the oil polluted the seaweed which is exposed at high tide and upon which sheep fed. As a result, thousands of sheep and six to seven thousand birds were lost in this catastrophe.

The year of 1979 was the worst year of all to date, as over 200,000 tons of oil were lost in the first five months; this almost equalled the total for 1977. This included a tanker loaded with twelve million gallons of crude oil that collided, on April 29, with an empty ship off the Brittany coast.

As early as 1957, a good study of an oil pollution spill site was made when the tanker *Tampico Maru* struck the coast near Ensenada, Mexico. Only 60,000 barrels of oil were involved. Fortunately, just before the accident, the site had been studied by the scientists at Scripps Institute of Oceanography. After the wreck, the shore was littered with piles of seaweed and dead and dying animals. The site was studied closely for recovery. Over the summer, algae and seaweed started to grow, but, during the winter storms, the remainder of the ship was torn apart and fragmented, the additional oil released destroying the plant life. As the water stabilized during the following spring, plant life and animal life were returning. It took years to normalize. Shelley said:

> Look on my works, ye mighty and despair!
> Nothing beside remains. Round the decay
> Of that colossal wreck, boundless and bare,
> The lone and level sands stretch far away.

I will mention only one of the other wrecks of some years ago. On March 18, 1967, the *Torrey Canyon*, loaded with 117,000 tons of Kuwait oil, was so deep in the water that it could only enter the harbor of Milford Haven at high tide. It was running at sixteen knots when it struck Seven Stone Reef at the Scilly Isles. The first slick moved towards the coasts of Cornwall and Guernsey, then turned toward the Breton Coast on April 10 and 11. The ship was bombed to remove it as a hazard, and the resulting second oil slick was much worse. It moved into the entrance of the English Channel, then to the south-southwest a hundred miles to the coast of Gascogne. This slick was fought at sea with detergents. It is difficult to say which was most dangerous, the oil or the detergents, since both damage sea life. I have a record of fourteen oil spills in 1981.

Much of marine oil pollution results from washing out the oil cargo tanks, then filling these with sea water as ballast. If sea water ballast was not used, the screws or propellers would be half out of the water and the ship would not be seaworthy. All tankers now are supposed to use the LOT, or Load on Top process. In this procedure, all dirty ballast and slops from tank washings are discharged into a slop tank on board ship and not run into the ocean. After the oil has floated to the surface, the water is drawn from the bottom until only a thin film of salt water remains. The fresh crude is loaded on top, and this the refineries can handle. Previously, two million tons of crude oil were washed into the sea; optimists now say that only one-fifth that amount is dumped annually. The amounts are too great even now. Many of the tankers run out to sea only thirty or forty miles and wash out the oil storage tanks.

Two dangerous sources of oil pollution, and one method of control, have not been mentioned. Only a few words will be used on each as the literature is voluminous.

First, oil pipelines can crack and break. This almost always occurs at the weld in arctic areas. It is difficult to weld in subzero temperatures, and, although there have been a number of breakages with relatively large oil losses in some instances, it is surprising that more have not occurred. The damage to the permafrost area is significant, due both to the direct kill of plants and to the screen formed causing peculiar absorption of sunlight and potholes in the permafrost.

Another loss of oil has been from oil drilling in the ocean. Some very severe oil losses have taken place in this way. Part of these occur because certain companies are too frugal to use sufficient casing to reach a safe depth; others have occurred because storms have snapped the pipeline or even the platform. In still other instances, spillage results because of carelessness and, perhaps, fire. Some may be due to corrosion of the casing and oil piping.

The Santa Barbara oil spill on January 28, 1969, was one that should not have happened. The casing did not go to a sufficient depth. A blowout occurred in which a large amount of damage was done to sea birds, seals, and smaller fauna and flora.

The last point that should be mentioned is that although the United States has started to prevent junk tankers, old and worn out, from landing on our shores, the penalties, including insurance claims, are not severe enough at present for complete oil spill cleanup.

* * * *

Poor man. From the vantage of 2376, we see how trivial were many of his concerns—transient, impermanent solutions to matters of physical discomfort. By the year 2000, man had acquired enough knowledge and skill to use substitute technologies instead of burning fossil fuels which had so many other, and more valuable, usages. But old patterns persist; and developing nations, with expanding populations, ate ever more quickly into precious reserves of oil, gas, and coal. As Cowper had warned:

> Our wasted oil unprofitably burns.
> Like hidden lamps in sepulchural urns.

The "Nukes" We Build

It may seem strange to say this, in light of all the adverse publicity nuclear power plants have received in the recent past, but—except for the breeder reactor plants—no death has resulted in all these years from the operation of such units of existing utilities. And this result has occurred despite some almost incredible stupidity in the design, construction, operation, and governmental controls applicable to at least some of such structures. Considering, also, that some utilities now derive half or more of all their electrical power from nuclear sources, under the constant threat of oil embargoes, it may be too late to turn the clock back, although we might have wished for sounder planning in the inception, and certainly will demand it from this point on.

Also, in introducing this subject, it might be mentioned that our navy operates more nuclear plants than all our utilities combined. Finally, as to nuclear wastes, only 1 percent of such wastes comes from utilities.

Now, with that off my chest, I must confess—like the chap who wouldn't want his sister to marry a cannibal—I'd just as soon not have such a plant as my next-door neighbor. But let's see how they tick, and whether nuclear energy fits into a plan for survival, or one for destruction.

The two most common types of nuclear reactors are the light water generating units, one of which is the boiling water reactor (BWR), and the other the pressurized water reactor (PWR). In either of these two types of reactors, the fuel, which is slightly enriched uranium oxide, is made into ceramic pellets which are inside of zirconium alloy cladding tubes to form fuel rods. The fuel rods are grouped together into fuel assemblies, and a number of these together make up the reactor core. Interspersed among the fuel assemblies are rods or gratings of cadmium, graphite, cobalt, or boron, which have the capacity to absorb neutrons; as long as these rods are pushed along the fuel rods they serve as dampers and there is no reaction. When these control rods are pulled all the way out, tremendous heat is given off.

What causes the heat is the fact that the nucleus or center of each atom is composed of protons and neutrons. We are ignoring the electrons, which are far from the core and do not participate in the reaction. All of the radioactive series of the uranium compounds have the capacity to give off either slow or fast neutrons, which can strike the nucleus of another atom of the radioactive series and split or "fission" the other atom, producing tremendous energy as heat and also giving off additional neutrons.

Thus, if the nucleus of highly radioactive uranium 235, which consists of 143 neutrons and ninety-two protons, is struck by a neutron, it will fly apart and can produce one fission fragment of barium 142, another of krypton 91, and three additional neutrons, plus heat. As the three additional neutrons strike other nuclei, the reaction runs faster and faster. If we control this reaction by building up a steady state or chain by use of cooling water which is converted into steam, and by the proper use of the control rods, the result will be a light water nuclear reactor. If we cannot control the rate of nuclear fission, the reaction continues like a wild fire and in seconds a nuclear explosion would occur.

All of the by-products of the controlled nuclear reaction are contained inside of the cladding rods. In addition to barium and krypton, radioactive iodine, yttrium, strontium, cesium, cobalt, tin, and other radioactive elements are produced. Over a period of two or three years, these "poison" the reactor rods so that they must be disassembled and purified. This is a dangerous process.

In the light water reactors mentioned, the BWR is the least complex. The reactor core, which consists of the fuel rods and control rods, is held in place in the reactor vessel. Water, which is the reactor cooling medium, is pumped in from the bottom. As it enters, the control rods are drawn out part way until the nuclear reaction begins. This is controlled so that the water is continuously converted into steam as it passes by the reactor core. The steam turns a turbine at high speed which, in turn, spins the generator, producing electricity. A secondary cooling coil condenses the steam, converting it back into water, so that this almost endless circuit can continue. The secondary coil is like a radiator. Whereas the original reactor steam is radioactive, there is no radioactivity in the cooling water unless a leak were to develop.

In the PWR, the basis is the same except the steam is heated to some hundreds of degrees. The radioactive steam of the primary loop gives up its heat to another secondary loop as in the BWR in which there is no contact. Here, however, the heat is sufficiently great to convert the water of the secondary loop into steam, which turns the turbine blades at high speed and spins the generator. This secondary coil is cooled by water. This actually inserts one more safety step in that there are two separate and isolated coils between the radioactive steam and the cooling water. In a plant producing 800,000 kilowatts of electricity, it requires a water flow of approximately 1,200,000 gallons of cooling water per minute for the cooling coils of the reactor.

There are approximately seventy-two nuclear reactors in the United States, depending upon which nuclear plants are operational at a given time. Accidents have plagued the plants of all, or almost all, designs so only a few of the scores of closures will be mentioned.

In 1974, the Federal Trade Commission charged the advertisements of the manufacturers of polyurethane to be false and misleading. Such material is widely used in insulation, yet it can burn almost like solidified gasoline. In that year, a major fire broke out in the Atomic Energy Commission's Fermi Laboratory near Chicago, the site of the world's largest particle accelerator. A workman burned off its polyurethane insulation with a torch; within eight minutes the 430-foot tunnel was engulfed in flames.

The Tennessee Valley Authority's Brown's Ferry nuclear power plant in Decatur, Alabama, almost had one of the worst disasters of the nuclear age. As we have discussed earlier, in 1975 a worker inspecting for leaks touched off a fire in the polyurethane foam insulation. As a result, the cables shorted out, rendering valves, backup cooling systems, and pumps unusable. Only after hours of fighting, the Athens, Alabama, fire department sprayed water on the cables against the advice of the TVA and extinguished the blaze. The losses could have been catastrophic.

It should be understood that atomic or nuclear explosions per se cannot occur in a nuclear fission power plant. A purity of uranium in excess of 90 percent is required for a nuclear bomb. Utility plants use uranium ordinarily at less than 5 percent. Therefore, one does not have to worry about a conventional nuclear plant suddenly disappearing in a mushroom cloud.

There are certain dangers, however, which should be recognized and guarded against. If the core-cooling system does not react properly the core could melt itself far down into the earth in what is known as the "China syndrome," where the nuclear material could disappear. Another type of disaster that could occur would be a steam explosion or hydrogen explosion, either of which, if sufficiently violent, could burst the outer container resulting in active fissionable material being carried into the air and dissipated to the surrounding area—causing carcinogenesis and death. The hydrogen at the Three Mile Island incident starting March 11, 1979, formed a thousand-cubic-foot bubble from the reaction of water with the zirconium cladding. At the extremely high temperatures the oxygen was probably converted to the metal oxide, thus preventing an explosion of the type mentioned. The first engineers reentered the contaminated reactor building May 20, 1980. Many causes were involved in that incident, such as inadequately trained personnel, defective design, and—it has been said—rushing of the reactor into operation with undue haste in order to secure a tax write-off. Jean Giraudoux stated, "There are truths which can kill a nation."

Next could be a loss of coolant in which sufficient water is not present due to the breakage of a large pipe or sudden depressurization of the water in the pressure chamber. The boiling water is not sufficient to cool the core of the reactor and even after a successful SCRAM the core heats up from radioactive afterheat. If the emergency cooling system reacts successfully, the reactor can be repaired.

The primary concern, however, is not the concept of "explosion" but the radiation leakage. This could occur through contaminated water permitted to escape; more probably, it would be fumes, or vapors, carried by air currents. There is no reason why, in the interests of safe operation, this is not constantly monitored having in mind, particularly, the direction of prevalent wind currents.

There are a great number of shutdowns of nuclear plants for various causes, frequently leakages, of which only a few of the recent ones will be mentioned for lack of space.

1. June 11, 1979—The Millstone II nuclear power reactor in Waterford, Connecticut, was shut down for about two days because of a leak of "mildly radioactive" water in a containment building.

2. June 19, 1979—Three operating nuclear power plants, including San Onofre, reported similar defects in a welded part that carries water to the reactor's steam-generating system. The staff of the Nuclear Regulatory Commission regarded the problem as a generic defect, one that might appear in any of the forty-three pressurized water reactors licensed in the United States, according to Edward Jordon, assistant director of the NRC's division of reactor operations in Bethesda, Maryland. The other two plants were the H. B. Robinson Nuclear Power Plant in Hartsville, South Carolina, where x-ray evidence showed cracks in nozzles that supply water to the reactor's steam generators, and the Donald C. Cook plant in Benton Harbor, Michigan.

3. June 30, 1979—In the fourth mishap in two weeks at the Peach Bottom, Pennsylvania, nuclear power station, radioactive steam escaped into the turbine building.

4. July 23, 1979—A Japanese nuclear power plant at Tokaimura was shut down when it developed a steam leak in the primary coolant system.

5. April 30, 1980—A small nuclear reactor discharged fifty times the permissible concentration of radioactive argon into the air within thirty feet of an air conditioning intake duct for a UCLA classroom. The amount of radioactivity that has entered the Math Sciences Building since it was opened in 1967 is unknown as it has never been monitored.

6. May 11, 1980—A leak in a coolant pump covered the floor of a reactor building at Unit 1 of the Arkansas Nuclear Power Plant at Russellville with eighteen inches of radioactive water. It is claimed that no radioactivity escaped as a result of this accident.

7. January 15, 1981—Radiation leakage occurred at Consolidated Edison's Indian Point Nuclear Power Plant into the Hudson River over a period of a month.

8. February 9, 1981—Two radiation leaks occurred within nineteen hours at the Fort St. Vrair Nuclear Power Plant near Plattville, Colorado, where helium coolant is used. This makes eight leakages at this facility.

9. April 21, 1981—A potentially hazardous leak of tons of nuclear waste was hushed up for forty days at the Tsuruga plant. This is the second accident of this type this year.

10. A large number of nuclear accidents have been detected both in the United States and abroad. As Adlai Stevenson said years ago:

> Nature is neutral. Man has wrested from nature the power to make the world a desert or to make the deserts bloom. There is no evil in the atom; only in men's souls.

May his words prove to be correct! One must remember, however, that for years the Atomic Energy Commission, predecessor of the NRC, deliberately withheld the dangers of nuclear reactors from the American public; and, at least up to the Three Mile Island reactor accident, the NRC had practically allowed all nuclear plants to monitor themselves with very little supervision. On January 25, 1979, the NRC repudiated a study it had used for five years to show that nuclear power plants are safe, and ordered its staff to review past and present licenses to see whether they were tainted by the Rasmussen report of 1974. On April 29, 1979, Congress was urged to demand tighter federal and state controls over nuclear power plants and to order more on-site assistance.

In retrospect, it is quite clear that this nation, in light of its abundant coal resources, should have moved more cautiously into the nuclear age. We should have built more, and enlarged, fossil fuel plants for perhaps fifty years, taking advantage of all available air pollution control devices. During that time, the nuclear reactors and such soft technologies as solar heating, thermoionic converters, thermoelectric converters, hydrogen-oxygen feed cells, and magnetohydrodynamics could have been studied. This period of time would not have been sufficient to have caused the "greenhouse effect" to melt the polar ice caps.

All studies of nuclear power systems should have been made in the laboratory until the system was perfected, then advanced to the pilot plant scale to make certain that there were no "bugs" in the system before building full-scale plants. We almost wiped out Detroit and several other areas because full-scale plants were built even though similar pilot plant models were acting erratically and indicating signs of danger.

Nevertheless, assuming we have the wisdom to learn from mistakes of the past, and that we will exercise better judgment in our design and approval of new plants, it seems that an enlightened judgment should be made. Are we, as a nation of people accustomed to adequate heat in homes and central buildings, air conditioning, and industrial power, willing to give up those conveniences, and possibly diminish our employment rolls? Or is the potential risk one which should be embraced—bearing in mind the possible "greenhouse effect" if only fossil fuels are used indefinitely for heat, air conditioning, and power?

Without nuclear power and vastly increased coal-fired plants our future is grim. As of February 1980, France had sixteen nuclear power plants versus the seventy-two in the United States. For the next five years, however, France will be opening a new nuclear power plant every two months and by the mid-1980's will be receiving 55 percent of its electricity and 20 percent of its total power from nuclear sources, whereas in the United States we receive approximately 11 percent of our electricity and less than 4 percent of our total energy from this source.

Let us make it clear that the potential risk is not acceptable, in the current state of knowledge, so far as "breeder reactors" are concerned. The experiences in the Fermi plant should fill us with caution. This was a large-scale plant operated by Detroit Edison Company and associates at Lagoona Beach, Michigan, which is relatively close to the town of Monroe, and perhaps more importantly, close to Detroit. The purpose of a breeder reactor, such as the above, is to pack uranium 238, which is relatively nonreactive, around the very reactive core containing uranium 235. The nucleus of the atom of uranium 238 would capture a neutron split off from the 235 and be converted to the highly reactive, highly poisonous, plutonium 239. Theoretically, the idea is sound, for in the United States we have little activated uranium. The difficulty is that sufficient work has not been done so that a safe reactor can be made.

The first experimental breeder reactor had been made in December 1951 at the Atomic Energy Commission's (AEC) reactor testing site near Idaho Falls. This reactor, EBR-1, produced sufficient power to light several 250-watt lamp bulbs. The Detroit reactor was based

more or less upon this. In December 1952, the Canadian government
had built a plant near Chalk River using heavy water, with which
U238 can be used. A mistake occurred and they had to dump most of
the heavy water along with regular light water that had become radioactive in keeping the uranium core cool. There was evidence of a
hydrogen-oxygen explosion inside the reactor, but, by good fortune,
the melted uranium was contained.

The Detroit Edison group applied for and obtained its construction permit from the AEC in 1956. In 1957, the University of Michigan
released the Gomberg Report, which stated that if a nuclear accident
occurred, and conditions were most unfavorable, approximately 130,000
persons might receive 450 RAD's (Radiation Absorbed Dose) and die,
and over 175,000 might receive 150 RAD's which could cause nausea and
possible leukemia.

Chalk River in Canada had started other tests in 1958, with almost
disastrous results. The controls were unreliable; and, on May 23, 1958,
one fuel rod had been badly damaged and two others showed a high
intensity of radiation and had to be removed. It was a dangerous and
almost deadly job. There was evidence of an explosion, and it took
months to decontaminate the building. If U-235 or P-239 had been in
use, everything for miles around would have been destroyed.

At Idaho Falls, reactor SL-1 (not the same as EBR-1 above mentioned), an experimental model using U-235, was having difficulty in
December 1960. It was shut down for a week, and on January 3, 1961,
started again. The three operators were having trouble with the cadmium
control rods jamming and sticking. The group alarm went off, but, by
the time the other members could reach the plant, the explosion in
which three men died had occurred. As Hilaire Belloc would have said:

> A smell of burning fills the startled air—
> The electrician is no longer there!

On August 6, 1966, Fermi reactor No. 1 at Lagoona Beach, Michigan, was started, but steam leaking in the generator occurred and the
plant was closed until October 4, 1966. The withdrawal of the control
rods was started that evening and all appeared well, so the reactor was
put on standby till the next morning. One steam generator valve did
not function properly, so it was early afternoon before the tests started.
Within an hour trouble occurred. Radiation was leaking, and they

could not find the trouble. Three subassemblies were bad, and it was fortunate that the plant was not destroyed. It was never placed into effective operation.

A gas-cooled fast breeder reactor (GC FBR), using helium as the gas and converting U-238 to P-239 or Thorium 232 to U-233, has not been used on any scale in the United States but has been used in the United Kingdom where carbon dioxide is the gas of preference.

In Britain, the Windscale reactor was different from that of the Fermi in that the neutrons were modified by graphite. The plant had run successfully for some years, and, upon occasion, the blowers would be shut down so that the temperature of the graphite block could rise and release the extra energy from the graphite. On one occasion, something went wrong. The cadmium control rods were shoved all of the way in, but the radiation in the giant stacks began to rise again. All of the district was monitored for radiation and the milk at the farms in the area was destroyed as it was high in radioactive iodine. Fortunately, no strontium was detected.

If breeder reactors are ever considered, much more work should be done than has been performed up until now. It would be a raging beast, once out of control. And, after designs have been studied critically, until a pilot plant—well removed from civilization—has been operated for a five-year period with no incidents, large or small, no approval should be given for a working plant. Even then, such plants should be placed at least fifty miles from any metropolitan area and where prevailing winds would not carry any airborne radiation to large population centers.

It is hoped that, even with more conservative water reactors, no losses will occur. But should disaster strike, certainly the victims should be adequately compensated. And, at present, they would not be. The Price-Anderson Act of Congress of 1957, renewed in 1965, set a $560 million liability limit for nuclear power accidents and provided for liability payments to be made through a combination of insurance company pools, $5 million contributions from each of the companies owning the nation's operative reactors, and the federal government.

Members of the insurance and nuclear power industries seem to agree with the view of Ambrose Kelly, general counsel for the American Mutual Reinsurance Company in Chicago and manager of the Mutual Atomic Emergency Reinsurance Pool. He said, at a recent annual meeting of the National Association of Insurance Commissioners, that,

"The chances of such an accident occurring are remote. Everything went wrong at Three Mile Island, and it was still a relatively minor accident."

Everything did not go wrong at Three Mile Island. Had the great hydrogen bubble exploded it could have blown solid debris over an area of several miles, and the radioactivity could have been responsible for 100,000 deaths and 150,000 cases of leukemia. If one were considering lives alone, this would have amounted to less than $6,000 each from the insurance fund. If lives and leukemia cases were considered, less than half of that amount would be available. It is necessary under the present circumstances to think in terms of many billions of dollars. The Price-Anderson liability limits may even be tested in the class action suits resulting from the Three Mile Island accident.

Perhaps of even more present concern than the design of light water reactors is the problem of disposing of nuclear waste materials. And there is no question but that this is a long-term problem. The toxicity or radioactivity of some of the radioactive wastes is such that they will remain dangerous for 200,000 years or more. Some of the radioactive products have a half-life of seconds, such as nitrogen 16; bromine 85 has a half-life of three minutes; iodine 131, eight days; cesium 137, twenty-seven years; strontium 90, twenty-eight years; carbon 14, 5600 years; uranium 234, 270,000 years.

But, before we blame this problem upon nuclear power plants, let us understand where most of the debris, or refuse, originates. A total of over 500,000 tons of highly radioactive wastes have accumulated from military uses as opposed to 5200 tons produced by the seventy-two nuclear reactors used in industry,* a ratio of 100 to 1!

Nevertheless, whenever these dangerous wastes accumulate—and whether in large or small quantities—some safe disposition must be made of them, or they will remain a time bomb whose ticking clock represents a risk every hour that passes. At the power plants, spent fuel rods are immersed in giant pools in which they can cool until such time as there are approved reprocessing plants. Military wastes are held in steel tanks at the Hanford Military Reservation in the state of Washington. Leakage from these tanks has been almost 500,000 gallons of radioactive material. At present, double-walled tanks are being made to prevent future accidents. At Hanford, a few years ago, where transuranic (TRU) wastes were being buried in relatively shallow trenches, enough

*U. S. News and World Report, May 7, 1979, p. 68.

plutonium migrated from one ditch to another to almost make a chain reaction possible.

There are few places remaining to bury nuclear wastes, and it is estimated that in 1990 there will be six to seven times the amount produced annually as at present. Low-level wastes have been buried at Maxey Flats, Kentucky, where it was found that the plutonium had migrated two miles from the burial site. Of the six original burial sites, two are now closed (West Valley, New York, and Maxey Flats, Kentucky); a third site at Sheffield, Illinois, is filled to capacity; three that are still open are the ones at Barnwell, South Carolina; Beatty, Nevada; and Hanford, Washington. The Department of Energy has fourteen waste disposal sites of its own.

The United States has not been anxious to reprocess spent fuel as we wish to keep the products out of the hands of saboteurs and hostile military powers. In France, the Cozema plant plans to remove all but 0.5 percent of plutonium from water, but three times that amount of Americum is left which decays to plutonium.

France has developed a large plant for reprocessing spent fuel rods from its own country as well as from Japan, West Germany, The Netherlands, Belgium, Sweden and Switzerland. The unused uranium 235 and plutonium are extracted from the spent fuel rods. At Marcoute, near Avignon, the waste is permitted to cool for five years after which it is mixed with borosilicate glass and hardened into a block. The glass is expected to resist corrosion and seepage. The scientists believe that all of the nuclear wastes of the next twenty years would form a solid glass cube of fifty-three feet.

Between 1946 and 1970, the United States buried thousands of tons of low level wastes, mostly from weapons, off the coasts of Maryland and California. As I look at the masses of nuclear wastes that are dangerous for generations to come, I am inclined to agree with General Omar Bradley, who stated, "We have grasped the mystery of the atom and rejected the Sermon on the Mount."

A very brief summary of a few situations shows the struggle that is going on:

1. January 26, 1979—The Carter Administration told Congress that states should be denied veto power over waste disposal sites within their boundaries.

2. February 16, 1979—A long-abandoned uranium waste disposal site emitting radiation in excess of federal standards was discovered

beneath land now occupied by a business site in Denver, Colorado. Officials of the Environmental Protection Agency announced that seven other similar plants may be within the city limits of Denver.

3. February 23, 1979—Nuclear shipment accidents increased from 1.2 a week to 1.9 weekly, according to the Critical Mass Energy Project.

4. March 11, 1979—Social obstacles, it was stated, may be tougher to overcome than technical ones in solving the problems of permanent, safe disposal of radioactive wastes from nuclear power plants, according to a report for the White House by the Interagency Review Group.

5. March 14, 1979—The new governor of South Carolina, Richard Riley, was fighting the storage of radioactive waste within his state.

6. April 16, 1979—Design work on a multimillion-dollar facility to solidify twenty-one million gallons of highly radioactive waste was announced by the Department of Energy. The government's Savannah River plant, near Aiken, South Carolina, has been given two million dollars for design studies. The DOE is considering burying the solid waste in either New Mexico or Nevada.

7. May 17, 1979—Nevada Governor Robert List ordered an indefinite halt to nuclear waste shipments from Southern California to a dumping site, as the result of an accident that sent radiation into the atmosphere. The material was improperly packaged in gypsum instead of concrete, and a fire resulted.

8. May 18, 1979—It was disclosed that stored nuclear weapons can accidentally emit radiation, and that the Pentagon should notify local authorities of atomic warheads kept in secret caches in their areas.

9. July 6, 1979—A hearing to consider legislation that would provide safe transportation of radioactive nuclear reactor fuel and wastes was scheduled by a Senate science subcommittee.

There are at least thirty other such articles concerning different states, which will not be listed. The purpose is to indicate that no one seems to know exactly where previous dumping sites are located, and no state wishes to have these located inside of its border.

A great number of suggestions, mostly foolish, have been made. Recently a series of articles suggested that one of the islands of the Solomon group be used as a burial site of nuclear wastes. Needless to say, this was protested, and rightly so, at once. This is an area where hurricanes frequently sweep across the islands, dumping water in the area that could wash anything into the sea as soon as the containers have been sufficiently broken down by time and attrition. Even more foolish is the idea of burial of the hot material in the Antarctic where melting ice would assure its entrance into the ocean.

Another suggestion has been the burial in salt caverns in Kansas, or in an area approximately twenty to thirty miles east of Carlsbad, New Mexico. Most of the salt areas have been drilled, and there are many holes through which water can enter the caverns. If water enters through drill holes, or by percolation through the various strata, a concentrated brine will form which will corrode the containers and permit the radioactive material to escape. Even if by chance there are no drill holes, and assuming no cracks in surrounding rock, which is highly unlikely, the salt contains at least 1 percent moisture which is attracted towards a source of heat. Thus, a brine would form around the heating containers. Furthermore, salt structures are almost always underlaid with sedimentary limestone, and the leachate could get into the limestone and migrate into the underground water table.

Consideration also has been given to placing the containers in intercontinental waters far from the continental shelves. This would guarantee that the ocean fish would become radioactive.

It should be admitted that the United States has no plan for the disposal of nuclear wastes. On April 13, 1980, an excellent picture appeared of a helicopter on top of a thirteen-inch thick concrete dome covering more than 100,000 cubic yards of radioactive soil on Runit Island in Eniwetok Atoll in the Pacific. This soil was collected from the sites of forty-three nuclear tests in the 1940's and 1950's, primarily at Bikini Island.

Scientific tests were supervised April 19, 1980, in which in a cavern northwest of Las Vegas a series of holes were drilled twenty-feet deep in granite, which were then lined with concrete and stainless steel. Canisters containing nuclear wastes will be placed in some and electric heaters that will produce the same heat will be placed in others. The entire experiment will be monitored with sensitive instruments that will detect cracking of the concrete or granite, and any resulting leakage.

Boeing Aerospace Company was awarded a $296,000 contract May 20, 1980, by NASA to study the disposal of radioactive wastes in space.

I have no magic wand to wave in solving this problem. The hazards of certain proposals have been indicated. It is too expensive to place all such wastes in interplanetary missiles and land them on Jupiter or Saturn; and, if a missile ever misfired, the resulting havoc on re-entry could be horrendous. It seems essential to encase long-life radioactive materials in lead shieldings and bury them deep in granite or basaltic caves not ordinarily subject to water penetration. This may require the construction, by boring and excavation, of such openings in remote areas of the Rocky Mountains, and there are such locations rather well removed from populated areas. Of course, if the population explosion continues, as discussed in Chapter 11, there may be no remote area on the face of the earth.

A sound proposal for the disposal of transuranic (TRU) radioactive waste was made by Isaac J. Winograd in June 1981. At Yucca Flat, Nevada, the Sedan Crater is one of those blasted out by the Atomic Energy Commission in 1962 "Plowshare" project. It has a depth of 98 meters, a crater diameter of 376 meters and a cubic capacity of over 5 million cubic meters. It is in a zone where the rainfall is less than five inches per year and the underground water table is at a depth of 580 meters. As a result all of the TRU wastes produced up to A.D. 2000 could be placed in this hole and an overlay of eight meters of clay or silt could cover this to prevent the interference of man.

Another point of danger that should be mentioned lies in the original mining of the uranium. On June 8, 1979, the TVA announced that houses located in northern Alabama and other areas embraced within the Tennessee Valley Authority had been built with a phosphate slag sold to block makers. This was now found to be emitting radon gas. Homes in which radioactive tailings have been used for building blocks also are dangerous to their occupants and neighbors. Approximately 14,000 residents of central Florida whose homes are built over abandoned phosphate mines have a 35 percent greater chance of dying of lung cancer than normal, the EPA has warned. This is due to the radon gas, which is carcinogenic, escaping from the mines.

On July 23, 1979, New Mexico officials stated that 100 million gallons of water had spilled from a dam of uranium tailings and extended fifty miles down the Puerco River into Arizona. The dam at the United Nuclear Corporation near Church Rock, prior to that time, had been considered well-built.

There has been a definite amount of deceit on the part of the AEC, according to the Los Angeles *Times* of June 20, 1979. Although the AEC was warned in 1948 that uranium miners faced grave danger from lung cancer, the government failed for almost twenty years to set health standards for those mines, a Senate subcommittee was told. The story of an epidemic of cancer around mines in Utah and Colorado was given both by scientists and by the widows of the miners. George Val Snow testified that of forty-two friends and relatives who had worked with him in the early 1950's in a uranium pit near Marsvale, Utah, twenty-two have died of lung cancer. Eisenbud, a scientist with the AEC, said he had urged the AEC to demand standards of ventilation and other safety standards, but that he had been rebuffed.

In that same article, it stated that, in the atmospheric nuclear explosion tests, the Department of Defense overruled the AEC and permitted troops to operate as close as 1.1 miles from ground zero in the early 1950's.

An article of April 20, 1979, stated that by 1955 the AEC, and possibly President Eisenhower, knew that the nuclear tests were dumping high levels of radioactive fallout on populated sections of western states. The hearing brought testimony from the widows, or widowers, of cancer victims who had been exposed to radiation levels up to 500 times that of the Three Mile Island accident. One man's wife and nine other members of his immediate family had died of cancer in the last two decades. A large series of articles were written in 1979 which indicate that both soldiers at the testing grounds for serial nuclear explosions, and civilians who lived near enough to the area so that the wind carried radioative fallout, died as a result.* A series of suits against the government have been and will be filed. On January 29, 1979, Dr. Robert C. Pendleton, Director of the University of Utah Radiological Health Department, stated that he had been pressured into stopping part of his research because his findings contradicted official statements concerning the safety of open-air atomic tests in Nevada during the 1950's and 1960's. Pendleton stated that they were bucking a large military-industrial complex which believed the weapons' tests were for the benefit of the country. In 1981–82 compensation has started.

The unfortunates who died from uranium mining, or from exposure to radioactive fallout, can never be compensated; but, if their near relatives are living, they should be compensated for the agony suffered by both. As Vannevar Bush has said:

> Fear cannot be banished, but it can be calm and without panic; and it can be mitigated by reason and evaluation.

*See, for example, "Grim Legacy of Nuclear Testing," *Reader's Digest*, August 1979, p. 102.

It will be noted that we have not even discussed background irradiation, the build-up of irradiation in fish, fowl, or animals. Those are, however, considerations to be recalled.

There is one further point that might be borne in mind. There have been a number of mysterious disappearances of uranium in the last twelve years. In 1968, a coastal freighter, loaded with more than 200 tons of uranium from Belgian-owned mines in Zaire, disappeared from November 16 to December 2, en route to Genoa, Italy. The freighter finally docked in Turkey. It was then empty. The ship was owned by Dan Aerbelo, an agent of the Israeli secret service. It is strongly suspected that Israel received 206 pounds, or more, of enriched uranium that could not be accounted for. Another 391 pounds of enriched uranium disappeared from the Nuclear Materials and Equipment Company of Apollo, Pennsylvania. This amount of material could make six nuclear weapons of the size used against Hiroshima in 1945. Mr. Shapiro, the owner of this plant, is believed to have been active in recruiting scientists to work in Israel.*

Let there be hope that nuclear warfare, which could eliminate one or more nations and bring great devastation to others, will not occur in the Middle East as a consequence of the destruction of the nuclear reactor of Iraq by Israel. The writing of Einstein, in 1945, unfortunately, has much truth in it:

> Since I do not foresee that atomic energy is to be a great boon for a long time, I have to say that for the present it is a menace. Perhaps it is well that it should be. It may intimidate the human race into bringing order into its international affairs, which, without the pressure of fear, it would not do.

*Los Angeles *Times*, April 29, 1979.

The Resources We Destroy

We become aware of our diminishing resources when we walk along any of our shorelines at intervals ten years apart. The hungry waters first nibble and chew out the sandy beaches, then suddenly take a large bite in which homes and roadways disappear. The ocean has an insatiable appetite, and if a pier or jetty is built to protect a certain small area of the coastline, the ocean snarls like a raging beast in the next storm and tears out sections somewhat farther away that were previously considered safe. The Pacific, Gulf, and Atlantic areas have been involved in this process in which even cities have been inundated and man has fought a battle with the sea which, at best, he has not won. As Shakespeare said:

> When I have seen the hungry ocean gain
> Advantage on the kingdom of the shore,
> And the firm soil was of the watery main,
> Increasing store with loss, and loss with store.

Safeguarding our shorelines is important, in all areas of this nation. The following table is considered reasonably current:

	Total Shorelines Miles	Significant Erosion Miles	Proportion of Shorelines
North Atlantic	8,620	7,460	87%
California	1,810	1,550	86%
Lower Mississippi	1,940	1,500	81%
Great Lakes	3,680	1,260	34%
South Atlantic Gulf	14,620	2,820	19%
Texas Gulf	2,500	360	14%
North Pacific	2,840	260	9%

A battle is in progress now, in the Los Angeles area, that would lead to the change of California coastal protection laws in which approximately nine miles of the Santa Monica Mountains would be exempted from the present laws and could be used for development. Another fight was in progress, in June of 1979, to prohibit Summa Corporation from developing the wetlands of Balboa Creek, which is the last marshland of the Los Angeles area. Both sides agree that this area is invaluable for the California least tern, Belding's savannah sparrow, crabs, and clams. It is a salt marsh that can never be replaced; whether it should be used for purposes of conservation or real estate development is the never-ending debate between conservation and "progress." A British conservation group, meanwhile, has purchased 5500 acres of marshland in northern England that otherwise would have been purchased by a Dutch group for farming. In California, slowly but surely, our coast lands are going from sanctuaries to real estate.

There is some reason to believe, even without the "greenhouse effect" elsewhere discussed, that the Greenland Ice Shelf and the Antarctic Ice Cap are melting and that the Atlantic is rising from six inches to a foot each century. Along all of our coastlines we should plant hardy grasses and shrubs to prevent the shifting of the sand.

Many protests have been made that the federal government does not own sufficient land for parks or other developments. Actually, the federally owned acreage increased from the 1950's to the 1960's, and in 1962 the federal government owned almost 771 million acres of land, or almost 40 percent of the United States. In that same year, it owned almost forty-five million acres or 45 percent of California, and 224 thousand acres or 0.7 percent of New York. In Montana and Nevada, the government owned 30 and 80 percent respectively. In the United States we have more than 2500 state parks with close to six million acres, not counting state-owned forest preserves. California has 171 with acreage over 700,000 and New York has 141 with acreage over 2.5 million. In the state of California we have 2100 municipal parks of 135,000 acres, which includes, in the city of Los Angeles, 167 parks with over 10,000 acres.

In other words, we do have enough land for recreational purposes if it is maintained and used properly.

There are arguments about the preservation of different areas. A few illustrations may suffice: On December 31, 1978, it was proposed to close the Clear Creek area of San Benito County to off-road motorcycles because this area contains asbestos deposits which have fiber lengths of the exact length and diameter to cause a lung cancer known as meso-thelioma; in February 1977, a park fifty miles in length was planned along the shores of the Blackstone River from Grafton, Massachusetts, to the Providence River in Rhode Island; in February 1979, logging was to be prohibited in fifteen million acres of federal wilderness lands, and in April a ten-man committee was established to recom-mend ways in which the U.S. Forest Service could improve manage-ment of the twenty million acres of land it administers in California; in June 1979, former Interior Secretary Cecil Andrus stated that a decision would be made as to whether 113 million acres should be considered to be eliminated from the wilderness areas (classification as a wilderness area eliminates new mining ventures, timber cutting, and other activ-ities that alter the landscape).

Many of our most vocal conservationists state that the wilderness areas should be closed. Do they mean by this that roads should not be cut through on a highway level, or that certain narrow roads for vehicular traffic should be permitted, or is the idea that the areas should be open only to people packing in on horseback or backpackers, or that these areas should actually be closed and no one allowed to enter? Many fiery arguments can be started at a meeting of conserva-tionists by asking this question, since most of the individuals have different ideas.

Timber is, of course, a resource of great value in our civilization. Unlike mineral deposits, trees are permanently renewable and replace-able—but not indestructible. There is no reason why mankind ever should lack for wood, pulp paper, and many other derivatives of our woodlands provided, first; that the demands for enlarged building areas do not result in the leveling of forests to provide sites for homes and other structures; and second, that care is used to prevent destruction by disease infestation, or by hostile fires.

Throughout the world, conservationists are staging reforestation and tree planting drives. At the end of 1978, the government of the Philippines estimated that between 200 and 400 thousand acres were destroyed annually by indiscriminate logging over the last ten years. The government has now banned logging in many critical areas and decreed that every Filipino ten years of age or older must plant one tree a month for five consecutive years. The government is also developing a 25,000-acre pine forest.

Children and groups have launched a tree-planting drive for the northern Africa area of Sahel so that the children of the future may have something else than desert sands. In March 1979, this drive was under way.

British Columbia fears a spruce shortage due to an error that has permitted overcutting of the land. In February 1979, the worst-hit area was the 7.5 million acre area of Fort Nelson. Rising wood prices as a result of this and other overcutting will affect rich and poor alike.

The United States Forest Service started burning brush firebreaks in the Angeles National Forest. They are called fuelbreaks since all of the vegetation is not cleared—merely enough to permit efficient fire-fighting.

It is essential that all areas that can, have grass, shrubs, and trees, for it is these that utilize the carbon dioxide we breathe out and which, in turn, liberate oxygen that we need so badly. We have learned in our forests to do intelligent burning, merely to burn the litter each year and not permit it to accumulate until every forest fire becomes a crown fire involving all of the trees. We only recently started this practice, so it will be a number of years until all of the woodlands are in shape.

Since timber is a valuable resource, as land is cut over, the area should be replanted to trees that would be of value to our industries. Present figures indicate a total of almost 488 million acres of commercial forest lands, and slightly over 252 million acres of noncommercial forest lands. With improving techniques of planting, and of tree selection, stronger trees are developed and brought to maturity much quicker than in earlier decades of "natural selection." Forest lands also utilize areas not desirable for more common forms of agriculture.

It is likely that minerals of value exist under many timberlands; and, as demand increases for such minerals, exploration may be undertaken. Already there has been drilling for oil in several such areas. It is quite possible for such programs to be carried out with minimum damage to the surface forests.

We are, of course, experiencing some shortages in mineral reserves, making conservation—and reclamation—of metallic materials important in future years. To mention only a few minerals, which in this nation we use fairly heavily, the following figures may be of interest.

Metal—U.S.A. Usage Pounds Per Capita		U.S.A. Reserves % World Total	World Supplies Will Last (Years)
Aluminum	17	Negligible	200
Chromium	20	No reserves	Over 200
Copper	26	18.4	62
Iron	1570	8.5	177
Lead	20	35.6	49
Magnesium	5	Unlimited	Unlimited
Manganese	11	Negligible	197
Nickel	2.1	Negligible	77
Tin	1.1	Negligible	44
Zinc	17	20.1	41

We import all of our diamonds from Africa, plus almost 100 percent of our manganese and 90 percent of our chromium and cobalt—which are strategic materials. We also import 80 percent of our tin and nickel. Much of this can be salvaged from our own scrap to make us less dependent upon the remainder of the world. Tons of slightly radioactive scrap metals are stored at nuclear fuel processing plants and are safe to use as reserve materials. Phosphate, which was lost from our soils by erosion, may be replaced since deposits covering hundreds of square miles have been found sixty miles off the coast of North Carolina. Manganese may also be recovered in nodules from the sea. All of us should share in the recycling of aluminum and other metals.

There are others, of course, but these are the principal ones to be considered. Zimbabwe is our principal supplier of chromium; tin comes largely from southeast Asia, the Congo, and Brazil. It is a fairly safe conjecture that, within twenty years, many of our minerals will be extracted from the ocean; within fifty years, many might be mined on the moon.

We also must be concerned with resources of the surface, in addition to timber. Much of our best agricultural land is being wrecked by the building of cities and towns in these areas, and the fashioning of tremendous freeways and turnpikes, so that the land is no longer a support for plants that furnish food to us but has become a concrete jungle. The farmer, himself, is also becoming a diminishing resource.

The United States is moving toward agri-industry, which means that the small farmer can no longer make a living. For years, families were able to farm eighty to 160 acres profitably. Then, as larger machinery was developed, 400 acres or more became a small farm. When the small farms are gone, and only agri-business remains, there well could be severe economic, as well as social, changes. Census statistics show that thirty-four million small farms have disappeared since 1935. Twenty-seven thousand went bankrupt in 1979. Many of these were absorbed by larger farms; others were purchased for sites of industry. Only 1.8 million farms remain. As one farmer stated, "Who else do you think would be dumb enough to make a living buying retail and selling wholesale?" Even the irrigation regulations of California are slanted toward agri-industry rather than to the small farmer they were designed to assist.

This means that marginal lands, which now can support a farm family which survives upon the crops and animals it produces, with only modest income derived above such survival, will go out of production. The life style is one factor which induces such persons, at present, to remain. And the combined yield of marginal lands can add up to an impressive percentage of total production. But, under assembly line methods, those areas will not be so utilized. They may, then, of course, be converted to timber farming, which could represent a more economic method of utilizing such lands.

A major threat to our natural environment exists in that body of men known as the Corps of Engineers of the United States Army. There is no river so pristine, so meandering, and so lovely that this group cannot destroy with a dam. They seem to find a natural habitat of this type, beloved by man and by wildlife, offensive to the engineering mind, which prefers straight lines and racing currents.* They are enamored of dams, perhaps because irrespective of cost to the taxpayer, these are gigantic monuments to which they can point with self-congratulation.

*See the more extensive criticism of this body in *The Midas Touch: Dynamics of Market Investment* (1976). The American Publisher, Box 102, Oxford, Indiana 47971.

The fact is that many dams are unneeded and even detrimental. As pointed out earlier in this book, the dams erected in certain impoverished nations may well destroy the crop soils on which they presently depend. In this country, little judgment has been used in selecting and erecting dams, in cutting canals, and in locating jetports. There is a serious question whether some dams, such as Oroville and Auburn, are even earthquake-proof. If it develops that they are not, may the Good Lord have mercy on those living downriver!

Of course, man's folly is not limited to the Corps of Engineers. Nowhere has this been more clearly demonstrated than by importing, without adequate consideration, plants, animals, birds, and insects not indigenous to this country, with regrettable consequences. Every such mistake can tip the prospects for survival adversely, or destroy the enjoyment of that environment. And we should use more care in the future than in the past.

The water hyacinth was imported into the United States from South America for a horticultural exhibit in 1884. Within three years, that plant was clogging up streams and lakes. By the 1960's it had been causing damage in excess of fifty million dollars annually, blocking irrigation ditches, canals, and other waterways. The sediment of dead plants can build up a foot a year. And control efforts, first using mechanical means, then the manatee, and now herbicides, have been more palliative than real.

The Klamath weed, introduced from Eurasia into the United States, took over millions of acres of range land; kudzu, introduced into the southeast to nourish the soils, has taken over as a rampant weed; the barberry, brought in for ornamental hedges, harbored wheat rust; and the fungus on the imported Chinese chestnut, which was resistant to it, destroyed the valuable American chestnut trees. But, to show no favorites are played, Americans introduced the prickly pear into Australia, where it began to take over thousands of acres.

Leopold Trouvelot, a French astronomer at Harvard, a little over a century ago wanted to create a resistant strain of silkworm, so he imported the gypsy moth. It causes untold millions of dollars of damage to evergreens annually. And, in South America, to increase the hardiness of domestic bees, killer bees were imported from Africa—with the current, highly publicized, results.

Mosquito fish were introduced into the United States to keep down mosquito larvae. They succeeded also in wiping out zooplankton, which feed on algae, causing algae to flourish wildly. The large freshwater snail, Marisa, was brought into Puerto Rico and the United States to destroy submerged weeds; it attacks rice, watercress, and water chestnuts as well. The walking catfish, imported from Africa, has been rapidly killing off game fish in a particular body of water, then strolling to a nearby area and continuing its predatory habits. The poison toads, which have escaped in Louisiana, have been able to kill domestic animals.

We talk, from time to time, of building a sea-level canal across Nicaragua. Let us bear in mind the experience of the Welland Canal, near Lake Erie. This permitted the lamprey eel to move into the Great Lakes, with resulting havoc to the fish therein. The sea level canal we propose would permit poisonous snakes from the Pacific to migrate into the Caribbean Sea with great harm to the resorts of those islands. It would also permit the interbreeding of closely related fish species which could not, in fact, reproduce and which would "mule" out.

One gentleman, some decades ago, thought it would be romantic to introduce the English sparrow and starling into the United States. He did. And these pernicious birds became pests in this nation, vastly depleting native species, multiplying endlessly, destroying crops, and, in the case of the starling, spreading disease. The rock pigeon, imported for squab as a fancied delicacy, became the pigeon flocks of the cities, with their droppings and filth spreading hepatitis, psittacosis, ornithosis, encephalitis, meningitis, and histoplasmosis. The monk parakeet, brought in from Bolivia and Argentina, has multiplied throughout twenty-five eastern states, from Maine to Alabama, ravaging fruit crops.

We are all familiar with the stories of how rabbits, introduced into Australia, multiplied because of the absence of predators into the tens of millions, destroying crops and ground cover and requiring roundups and mass slaughter to maintain even partial control. Finally, disease controls are being employed. Sportsmen wanting a game animal introduced red deer into New Zealand; they multiplied so rapidly that they compete, with sheep, for forage and there is now a permanent open season upon them.

Muskrats were brought into Germany for their furs. Again, they multiplied and burrowed into canal banks and dams; even as nutria, introduced into Louisiana, escaped and dug into levees, causing leakages and flooding. Ships, in years past, settled goats on islands in their passage to assure supplies of meat and milk. They multiplied to the point where all forage, and ground cover, on those islands were destroyed, turning them into barren deserts; and both domestic wildlife, and goats, perished in most instances. The mongoose, brought into the West Indies to control rats, turned to birds and amphibians when the rat supply shrank.

When rabbits were brought onto some of the islands south and east of Australia, the result sounded like the "House That Jack Built," from the nursery rhymes. As the rabbits multiplied, cats were brought in to control them. This they did to some extent; then, they also multiplied, killing seabirds and eating their eggs—items important to the welfare of the islanders. Dogs were brought in to control the cats; they ate the seals. Next, one might suppose, tigers will be brought in to control the dogs, then white hunters to control the tigers.

Wisdom is needed before playing God in a state of nature. Certain flora and fauna flourish in a given area, but the balance of nature may easily be upset by the introduction of new species not indigenous to that nation. Pestiferous insects or plant life, in particular, may be introduced by accident, and constant vigilance must be exercised. But to introduce new species into either this nation or other lands may be folly indeed, as the few illustrations previously given demonstrate. If we are to preserve existing resources, such mistakes should not be repeated nor compounded.

It is well, of course, for efforts to be made at preserving desirable animal, bird, and marine life. This is not a problem solely of this nation, but of the entire earth. Unlimited fishing has almost produced wars over codfish; other nations have prohibited catches of tuna within offshore boundaries which have been stretched farther and farther, into the open sea—now 150 miles, in most cases. The United States and Canada have settled peacefully their questions regarding shore line limits.

But supplies of fish have been ravaged by factory ships, particularly those of Russia and Japan. New England lobster almost became obsolete with little heed being paid to the size and sex of those taken. The spawning of salmon has been endangered—in part by man-made obstacles, in part by water pollution, but in great measure by overfishing. The giant trawling nets for tuna have not only captured them in quantity, but also dolphin as well—and this is a source of ongoing controversy.

The United States has grown more strict in its measures to preserve fish. Our ships are overhauling trawlers, primarily Japanese; and, if the catch has been exceeded, or if they have been fishing within our territorial waters, the ships are brought into port, the catch is confiscated, and the ship is held until a fine is levied. Frequently, the same ship is repeatedly caught for this offense.

In 1978, Japan accepted the whale quota catch set by the International Whaling Commission for the 1978–79 season. The sperm whale catch had been cut 41 percent from the previous year. Although, at first, Japan and Russia insisted upon 6444 for 1978–79, they finally agreed to 3800. In June 1979, the United States proposed an indefinite moratorium on commercial whaling to permit management. In our own Glacier Bay in Alaska, sight-seeing ships looking for whales must slow to ten knots and stay a quarter of a mile away from all whales. Japan remains the country that will not agree with quotas on whaling and in mid-1981 has insisted on increased whale kills. Their kill of porpoise is also high and it is dubious if damage to nets is a justification. In many instances it appears to be meat hunting.

There is not sufficient time to discuss the kill of baby seals in Canada, the destruction of sea otters on the California coast, or of innumerable other species that are being destroyed throughout the world.

Since all of these comprise resources of mankind, it is important to preserve them even as we must preserve tillable soils and potable waters.

The general subject of conservation is not within the province of this book. We are becoming more acutely aware of the dangers to animal life and bird life from poisonous sprays, and this is well. Without a flourishing and substantial bird population, in particular, food crops and timber will be ravaged by insects which will play havoc with them. Farmers would be well advised to maintain hedge rows, instead of barbed wire fences; the few acres of cropland so sacrificed will be well rewarded by the natural controls nesting in those shelters. Multiflora hedges provide food as well as shelter, although they can become pestiferous in volunteer plants springing up elsewhere from seeds dropped by the birds.

In California, insofar as birds are concerned, the year 1979 was very interesting. Earlier it was mentioned that the level of Lake Mono had fallen so greatly that the island where the gulls nested had become attached to the mainland by a land bridge over which coyotes, foxes, and other animals could cross and raid the nests or catch the birds. The California National Guard, with a slogan, "We're all for the birds," blasted a deep trench through this peninsula so that the birds are temporarily safe.

Man is inextricably intertwined with his environment. Not only are the resources of which he can make immediate use important to his survival, but also every living thing must be taken into consideration in the pattern of existence. As Stewart Udall once stated:

> A land ethic for tomorrow should be as honest as Thoreau's *Walden*, and as comprehensive as the sensitive science of ecology. It should stress the oneness of our resources and the "live-and-help-live" logic of the great chain of life. If, in our haste to "progress" the economics of ecology are disregarded, by citizens and policy makers alike, the result will be an ugly America.

Our Exploding Populations

The final evolvement of present man, who calls himself *Homo sapiens*, is from a long line of primitive hominoids, or manlike creatures, in Africa. From some primitive form, perhaps the one known as Proconsul at the border of the Pliocene-Pleistocene epoch of some millions of years ago, one emerged that we can vaguely recognize as the ancestor of the present race of man. José Ortega y Gasset has said, "Primitive man is by definition tactile man."

This could refer to *Homo habilis*, who lived approximately one and a half million years ago, or even to one of his predecessors. He was a social animal who lived in semipermanent camps and had developed a food gathering-and-sharing economy in addition to making crude weapons or tools of stone. *Homo erectus*, who followed him by some 500,000 years, also walked upright, was more highly developed, and formed a culture more advanced than the split-pebble type of his predecessor. He had one great advantage over his predecessors, namely, the lack of a hairy coat. Since his hunting was in daylight under the hot African sun, there was no heavy, wet covering to slow him down as his metabolic activity increased. He was a more successful hunter, although he had to develop a very dark skin to protect himself against the sunlight.

Homo erectus was the true naked ape who had learned to survive. As time passed, he carried a sharpened stick of which the tip later was often fire-hardened, after he had learned to use fire. He learned to carry a flint flaked to a cutting edge or bound to a heavy bough, giving him weapons that would pierce, scrape, or cut.

A portion of the group began migrations northward, and, in the less intense heat than that of the tropics, animal skins were undoubtedly scraped and used as clothing. As they migrated northward, the groups or clans divided—some moving toward the north and east into the Orient, and others into Europe. The art of fire-making was first developed by the clans or tribes moving to the Orient probably 400,000 years ago, then somewhat later by those in Europe, as shown in the case of *Homo erectus*.

Mutations were taking place and *Homo erectus* had developed mutants that were *Homo sapiens*. In Europe, and undoubtedly in Asia, the interbreeding of the stocks produced modifications. Neanderthal man of 100,000 years ago is usually classified as *Homo sapiens*. His camping sites are easily recognized by the Mousterian chipped-flint stone axes he carried.

The peoples of Asia also passed through a series of changes. It is from this group that *Homo sapiens* came to North America by means of the Behring land bridge. They would be classified probably as early mongoloids, since the loss of brow ridges and the flattening of the face of later mongoloids had not occurred. Like most of the other groups, they were hunters who also collected roots, nuts, and berries. They were following the paths of the bison, camels, horses, and other animals that preceded them across the land bridge.

There were several periods within the Wisconsin glacial age when sufficient land was exposed to make a wide bridge, for the depth of the sea fell as the great masses of ice accumulated on earth. Many of the animals and man probably crossed as early as 60,000 years ago, but it was probably about 20,000 years ago that the tribes called the projectile-tip people crossed, following the wild animals. Their sharp flint points were attached to spears and they had learned to hunt together as a group.

In Los Angeles, the Rancho La Brea Tarpit Museum shows the types of animals and relative numbers trapped from approximately

40,000 to 10,000 years ago. Most of the large animals could have fallen prey to man. Pleistocene man was not a conservationist. In Africa, which had been his first home, the massacre of animals or what is known as Pleistocene overkill occurred 45,000 to 50,000 years ago; and, as a result, more than a third of the species known at that time were destroyed. In North America, Pleistocene overkill took place later as man had arrived late, so that most of this occurred 13,000 to 14,000 years ago.

Although *Homo sapiens* in most of these areas had become a skilled hunter and dug pits along game trails to trap animals, he found that most of the danger could be avoided by using fiery torches. With these, the clan or tribal group could stampede the larger animals over cliffs or drive them into swamps where they were easily killed.

Quarles stated, "It is the lot of man but once to die." Early man did not plan to die if he could escape by cunning. As late as the 1800's, the Indians of the Great Plains stampeded bison over cliffs, killing hundreds in a single drive and obtaining the hides and meat for a long period. They were not conservationists either, as they cut out the tongues and only chose the better parts of the bison in years when hunting was good.

In the migrations of early man to the north, he had to learn many things. Since he was hairless, protection against both the sun and against the cold were necessary. If he had too little Vitamin D, he would develop rickets; if too much, a condition that we know as hypervitamonosis D, which disturbs the calcium and phosphorus metabolism, could occur, resulting in death. The use of animal skins as a partial cover and a varied diet permitted man to migrate northward to live, although possibly not in comfort.

Migrations into Australia by land may have occurred at approximately the same time as migrations into the Orient. It seems likely that man moved from island to island which lay close together, for in the glacial age many land changes were taking place. In North America, the Cascade and Sierra Nevada mountain chains were just being elevated and this may well have been coupled with changes in the ocean. There are people who believe that at one time a great island or subcontinent called Mu existed in the Pacific Ocean and sank approximately 30,000 years ago, due in part to the recession of the ice sheets, volcanic action, and gross movement of the tectonic plates. Whatever happened, an isolated area (Australia) remained with flora and fauna that disappeared from other areas.

During these early years, man was a nomad and could do little to injure the land as he was living on the animal and vegetable materials produced by nature. He probably became associated with the dog during this time as a mutual benefit in hunting resulted. His migrations had placed him in Europe and in Asia 400,000 to 500,000 years ago but his life span was short. Although he had improved his armament by developing the throwing stick, which improved the utility of thrown rocks and spears manyfold, he was still one of the hunted as well as the hunter. As a result, there was no great increase in population. The average age of Neanderthal man was probably somewhat under thirty years, which may have increased to slightly over thirty years until the Mesolithic period of 10,000 years ago. There is no way to state precisely how many persons were on earth at this time but probably an estimate of five million would not be too far wrong.

Had man in the Mesolithic period reproduced at the rate he is multiplying at the present time, there would not only be billions, but trillions of persons on this planet, assuming that there was food and space for all. This is easily illustrated by what is known as the "J" curve, and the folding and refolding of a theoretical sheet of paper one millimeter (1/25 of an inch) will illustrate the point. If we folded the paper once it would be only two millimeters thick, if we refolded it twelve times it would be one foot in thickness, if we refolded it twenty times it would be 256 feet in thickness, at the thirty-seventh fold it would be the distance from Los Angeles to New York City, at the forty-third fold the distance of the earth to the moon, and at the fifty-second the distance of the earth to the sun.

If this had happened with the human race, and each generation time was twenty years, at the end of fifty-two generations almost innumerable people would be populating this planet. One thing that has prevented this, of course, is that the life span of man did not approach forty years of age until the Middle Ages; and, even through most of that period, the life span was relatively static. Homer sang, "By their own follies they perished, the fools."

There was more than this involved in that it was not until approximately six thousand years ago that man was practicing village farming to any degree. Before this, he had followed the herds of cattle and sheep as a nomad. Then he realized that, with the help of the dog, he could domesticate these animals, confining them to a given range. He had begun gathering the seeds of the cereals growing in the vicinity and eventually had planted these, first using a sharpened stick to punch holes in the soil and later a stick with a fire-hardened point to produce a furrow. In the Bronze Age, he produced a satisfactory plow.

Warfare between the villages—and, later, as the civilizations of Ur, Chaldea, Babylon, Egypt, and Sumeria flourished, between nations— took its toll of the populations. As Greece and Rome developed, the life span of man was still slightly above thirty years, due to war and pestilence. It is thought that by A.D. 0, the population of the world was approximately 250 million persons.

From this 250 million, it appears that man doubled his population to 500 million in slightly less than 1700 years. Normally, the date of A.D. 1650 is given. He doubled the population again by 1850 to one billion, this doubling requiring only 200 years. In eighty years, in 1930 the population again had doubled to two billion persons, and in 1975 he had again doubled the population to four billion persons. This last push took only forty-five years. At the present rate of reproduction, in A.D. 2010 the population will be eight billion, and in A.D. 2040 it will be sixteen billion. The last periods of doubling each will be in thirty years or less.

Recent statistics indicate that the doubling rate of population for certain countries is as follows:

Costa Rica	19 years	Guinea	31 years
Mexico	21 years	Uruguay	58 years
Pakistan	21 years	Japan	63 years
India	27 years	USA	70 years
Dahomey	27 years	UK	140 years

The rate for the United States is more aggravated than that shown, since illegal immigration is flooding in from all over the world, but principally from Mexico. It is rather a grim fact that the East Germans have planted one million land mines, 34,800 self-firing machine guns, and regular patrols with 1000 watch dogs to keep their population from escaping. At the same time, many Americans demand that a fence be maintained along the Mexico-USA border to keep out illegal immigrants. On June 26, 1979, President José Lopez Portillo announced a plan to crack down on the large-scale smuggling of illegal aliens into the United States.

Mexico, which had a population of sixty-one million in 1978, is expected to reach eighty-three million by 1985, and 132 million by the year

2000. This means that the government of Mexico must find new employment for 750,000 each year, merely to stay even with its present unemployment rate of over 30 percent.

The United States cannot hope to absorb an unchecked immigration tide since many or most of those coming to the United States will have, or will produce, large families. One method of checking upon the legal status of those in this nation is, of course, to require each individual to carry identification that cannot be forged and may be produced at any time. This sounds like a police-state tactic but it is not. When I lived in the United Kingdom, I had to carry my identification at all times and this was not easily procurable, except legally.

The persons most injured by illegal immigration are the Americans of Mexican descent, since they are the first to lose their jobs to the illegals. At times they might be sympathetic to those entering illegally, but, as social services to them are necessarily cut in order to support the others, much of this sympathy has been lost. As a result of the current taxpayers' revolt, it is probable that welfare will be "cut to the bone." It is to the advantage of all American citizens, except businessmen and ranchers who wish cheap labor, to stop the influx of illegal labor. The solution may lie, in part, in penalizing the employer who encourages the illicit traffic in needy aliens.

The problem should be attacked, both in Mexico and in the United States, through limiting the size of the family. This will be discussed later. As Joseph Priestley said, which is still true:

> Becky Sharp's acute remark that it is not difficult to be virtuous on ten thousand a year has its application to nature; and it is futile to expect a hungry and squalid population to be anything but violent and gross.

In Brazil, according to *Time* magazine of September 11, 1978, in spite of one of the world's most spectacular rates of economic growth, sixteen million youngsters are hopelessly deprived. The outcasts are called "nobody's children," and vary in age from infants to teenagers. In Rio de Janeiro alone, a hundred children under three months of age are abandoned each month. Brazil spends only $38 million per year on children's services, and this is poorly distributed. There is only one government, or private, agency for every 10,000 needy or abandoned children. Those who fall into the hands of the police, according to this article, are subject to beatings and rape. In such an environment, these wolf packs, in time, may tear down the present government. In Peru, it was estimated in 1970 that hunger killed a child every eight minutes.

Under the Reform Decree of March 1980, all farms of El Salvador of 1200 acres or over were seized and divided into tracts of five acres, which were to be turned over to the *campesinos* to work as their own property. In 1981 all farms between 240 to 1200 acres were to be divided. The fighting involved in the bloody revolution in progress between government soldiers, guerillas, and death squads has kept the program from progressing.

In Africa, the position is difficult to understand. Kenya has expressed the desire for fertility clinics rather than birth-control clinics. The population of all countries of Africa, except Zimbabwe and South Africa, is increasing faster than the food supply. Zaire, which was a food exporter at the time of its independence, now must import half of its food supply, as must Zambia. Many of the African nations are badly governed, but even in those with enlightened leadership, none are working for population control.

In India, the birth-rate target is three years behind. This was blamed on the attempted forced sterilization program of former prime minister Indira Gandhi. In 1977–78, less than 700,000 men had been vasectomized, which was 30 percent less than the number of vasectomies in Madyha Pradest state alone in 1976–77. A number of persons previously vasectomized are attempting to have the operation reversed. In April 1979, Dr. P. L. Soni stated that in the previous three years of so-called family planning, police came into the villages looking for men to vasectomize. The methods were ruthless and in some towns people were killed in the fighting. With the fall of Indira Gandhi the program was largely pushed aside. What will happen now no one knows, since she is again in power.

In an interview with a reporter from *U.S. News and World Report*, June 19, 1978, Prime Minister Desai of India stated that he was against forced population control. He said that when the people are convinced that they will suffer and their children will suffer if they have too large a family, they will practice family planning. At the same time, he also stated that nuclear energy should be available in India and that it would never be used for nuclear weapons.

In May 1981, both Rami Chhebra of the Family Planning Foundation and Dr. V. A. Panandiker admit the hopelessness of India's population position. Twelve million more persons exist than were predicted and by the year 2000 the population will be more than one billion. Calcutta in 1981 still has 3000 official slums, some of which have only one to ten faucets, many of which do not run all day. In Darapara, which has ten faucets, more will be added. There is now only one faucet for every 20,000 residents. There is no electricity and disease is rampant.

The untouchables of India are still barred from tea stalls, barber shops, and similar areas in 1981, except for the 100 elected to Parliament. A number of the young men of the 100 million untouchables have formed a group known as Dalit Panthers based upon their study of the American Black Power groups. Bapurao Pakhiddey, a forty-year-old lawyer, defined the new militancy as "the courage to take those rights we have already been granted."

As a side comment, it might be mentioned that India not only is overpopulated, but also has already exploded an atom bomb and is building a four-stage rocket.

On January 8, 1979, most of the Asian nations were refusing to accept any additional refugees from communist-ruled Vietnam, Cambodia, and Laos. International refugee officials predict that as many as a million more refugees will attempt to follow the hundreds of thousands who have already fled to neighboring countries and to the West.

In China, according to a report of March 17, 1980, thousands of women have been seized and sterilized. The goal of the country is "one family, one child." The two-child family is a luxury that China can no longer afford.

In Bangladesh, Ziaur Rahman has been trying to solve the problem of food scarcity and rapid growth of population. It was only through the gift of over two million tons of grain, much from the United States, that complete disaster was averted.

On July 23, 1979, according to the Los Angeles *Times*, deep concern was expressed about the public health problems being brought into California by the refugee "boat people" and by illegal Mexican and Latin American aliens. Los Angeles County has no indigenous leprosy, but in 1978, 255 cases were reported to county health officials. Malaria cases have been doubling for the last three years, with the majority of the cases involving recent immigrants from underdeveloped countries. The number of cases of tuberculosis reported in 1977 was 1478, which is 10 percent over that of 1970; and this is a disease which had been almost eradicated in the 1950's. There are approximately

250,000 Indochinese refugees in this nation, and they have been arriving at the rate of 14,000 per month. In Orange County, Asian refugees have had an incidence of tuberculosis thirteen times higher than the indigenous population, and rates of some gastrointestinal disorders are ten times higher.

The World Bank estimates that the earth's population figures will climb to 6.3 billion by the year 2000, reach eight billion by the year 2020, and stabilize at perhaps eleven billion a few decades later. This is, I believe, overly conservative. But, even if accurate, most of this explosive growth will be in countries that cannot supply the bare necessities for the people already present. It is estimated that in twenty-five years Africa will have to accommodate an additional 535 million, Latin America 440 million, and Asia 2.1 billion. Population growth rates in the world average 2.3 percent a year, in Mexico over 3.0 percent, and in the United States 0.8 percent. John G. Gilligan, administrator of the U.S. Agency for International Development, believes that it would be almost impossible for the West to make up the global food deficit of 100 million tons per year which will occur in 1985, even if the West wished to do so. About 200 years ago, Malthus had written:

> Population, when unchecked, increases in a geometrical ratio. Subsistence increases only in an arithmetical ratio. A slight acquaintance with numbers will show the immensity of the first power in comparison of the second.

Malthus was predicting famine within his own times, but the great frontiers of Canada, the United States, and Australia served as food-producing areas so that man did not starve. We have no great frontiers remaining. We are faced with a simple equation, which when applied to the United States becomes:

Population change = Birth rate − Death rate ± Immigration.

On the third planet, Earth, 250,000 persons are added per day, which is 1.7 million per week, or ninety million per year. The greatest difficulty is that the percentage of population under fifteen years of age is highest in the most economically deprived countries having the greatest birth rates. This means that there are more women in the age group of fifteen to forty years to bear ever more children, and the rates will skyrocket even higher than at present.

In the United States, there were four million people at the close of the Revolutionary War, thirty million at the end of the Civil War, seventy-five million at the beginning of the twentieth century, almost 100 million at the start of World War I, 140 million at the start of World War II, and approximately 226 million in 1981.

There was a "baby boom" after World War II in the United States. Before the war the fertility or fecundity rates varied between 2.2 and 2.5 for the preceding five-year period. This is the number of children per mother or family. After the return of the servicemen, the rates beginning with 1945 were, respectively, 3.0, 3.3, 3.7, and 3.4. It then began to drop and is now close to 2.2 or 2.3. A rate of 2.11 gives what is known as a "population zero" increase as it allows for the percentage of deaths that would occur if every family had only two children. A difficulty arises in that a large number of babies born in the postwar years are now mothers. It appears as if the number will average out at 2.3 or somewhat higher, although there is pressure on some groups to have more children.

In the United States in 1977, there were 462,000 legal immigrants, or approximately 1265 per day. President Carter at that time announced that we would open our gates to a very large number of immigrants from the Orient who in 1979 were known as the "boat people." In addition, we have approximately seven to nine million illegal immigrants, as stated earlier. A larger number of persons in the future will be faced with diminished support, because not only will the working population decrease because of the rising age of the population, but also many will tend to drop from the labor force to welfare.

The U.S. Census Bureau estimated, based upon the fact there were 209 million persons in the United States in 1972, what the resulting populations in the year 2020 would be if the average children per woman were as follows:

Average Children Per Woman	Population of USA in Year 2020
3.10	447,000,000
2.78	397,000,000
2.45	351,000,000
2.11	307,000,000
2.11*	280,000,000

*With no immigration.

There is more and more pressure to educate each child in the family, whether this education be through college degrees or in skilled technical fields. The cost to raise and educate each child, in the United States, varies between $15,000 to $50,000. In addition there is the desire in each family to have possessions, whether these include a good car, television, household appliances, or other symbols of status or labor-saving devices. This, in itself, will appeal to many families as a reason to limit the size of families.

If population is permitted to go unchecked, we encounter two interesting situations. First of all, at a given population level—and this will depend upon the wealth of that particular country—we encounter a situation where the population must level off immediately with no increases to maintain a "quality life." This is where persons are well fed and have relatively good shelter, clothing, and economic advantages. If we pass this point, with numbers continuing to increase, we then reach a number at which the population must level off immediately to have minimum survival. If this point is passed, and the population still increases, we reach what is known as a "dieback" curve. Persons, usually children, will starve to death until the point of minimum survival is reached. This has happened in certain areas of Africa and the Orient and presents a warning of the future for the United States or any other nation. This growth can be depicted as a large lily which has had suitable nourishment until the point when the blossom is opening, at which time suddenly it is deprived of food and water. The lily dies and only a stench remains.

It is essential that we control population by some method. Speaking as a biologist, and disregarding for the moment religious proddings, I recognize the fact that mankind must strive for and attain the so-called "population zero," or no more than 2.11 children per family. This does not mean per woman, for the man is included also. If one were to use the most stringent methods, after a family has had two children the male would have a vasectomy and the female would have tubular ligation, neither of which would interfere with the sexual enjoyment of a marriage.*

As I wrote the above, I had just reread an article of June 21, 1979, concerning Uruguay, a strongly Catholic country in which 150,000 abortions occur every year. Uruguay has been a very advanced social

*Special arrangements might be made, in given instances, for a third child.

state, but in January 1980 the church and the military came out strongly against abortions.

Many are firm believers in abortions where rape or incest have occurred, or the health or life of the mother or the fetus is threatened. It is also well known that if certain radioactivity levels are reached, the fetus will be irreparably damaged. In these cases, or in such situations as deformities caused by the use of a toxic drug such as thalidomide, it should be within the right of the mother, only, to decide whether she wishes an abortion. The reason this is stated so strongly is that there have been cases in which a man refuses to marry a woman with whom he has had intercourse, but still has gone to court to prevent her from having an abortion. And many persons not only oppose abortion but birth control as well.

Contraceptive devices are undoubtedly better than abortion. There is a long history concerning these. In an Egyptian papyrus of the middle kingdom, around 1850 B.C., a pessary made in part of crocodile dung was recommended. It has been interesting to try to determine theoretically if this had any scientific value, possibly through the presence of putrefactive amines which might destroy sperm cells; whether it was purely superstition, or just hastened early withdrawal, is not known. In another papyrus of 1550 B.C., the use of fermenting acacia tips for lint tampons is mentioned. Here lactic acid would have been formed, and this could be effective. Other early contraceptives included honey and sodium carbonate (an alkali) that could be spermicidal; preparations of oxgall, which is a surface active agent; and preparations of pitch, which contains phenols and cresols that are definitely disinfectants.

Aristotle mentions the use of oil of cedar, ointments of lead, or frankincense mixed with olive oil. The first two have definite active factors. Soranus, a Greek physician in Rome in the second century A.D., recommended occlusive pessaries, which were vaginal plugs of wool and gummy substances, and also the use of astringent solutions. Most of the other recommendations were sheer superstitious nonsense, such as the Roman women wearing amulets of asparagus, or charms.

Sterilization of the male and female is by far the most effective method of avoiding conception. The least effective are spermicides, the rhythm method, or withdrawal, in which cases the pregnancy rates per 100 women per year of exposure is twenty-five. The users are frequently called "parents." For the condom and diaphragm the rate is fifteen per hundred annually. If jellies are used with these, the rate declines. In the rhythm method, with a woman having a menstrual cycle of 28 days, the most fertile time normally is considered that period between the tenth and seventeenth days of the menstrual cycle. While the remainder of the time is considered fairly "safe" for avoiding conception, this is not always accurate. It is, however, the theory on which the rhythm method is based. Of course, the cycle can vary from twenty-one to thirty-five days.

The two methods most commonly used at present are the IUD (intrauterine device) and oral contraceptives, both of which have quite good records in prevention of conception. But the oral contraceptives have certain side effects that offer a risk factor.

The IUD's include the Lippes loop, the gold wishbone pessary, the Grafenberg silver ring, and one or two others. Each of these is inserted by a physician and has a lesser percentage of failures than does the condom or diaphragm.

The most effective method is the use of oral contraceptives, or "the pill." Since many exist, the so-called combined preparation of an estrogen and a progestin which is similar in chemical structure to two natural hormones, estradiol and progesterone, is usually recommended. They are not the same and are called "analogs," which means having a similar structure. When taken correctly, they inhibit ovulation, act on the cervix to prevent sperm penetration, and act on the uterus to maintain it in a state highly unfavorable for the implantation of a fertilized ovum. The technical details will not be discussed. It should be understood, however, that a risk is involved. Although risk factors may be low for the individual woman, thromboembolism, or blood clots, can occur and women with high blood cholesterol, hypertension, or a history of heavy smoking have an increased risk. They also tend to develop diabetes in middle age more frequently than those not on the pill. Other difficulties may occur, and the use of the pill should be thoroughly discussed with the physician.

A remark of Adlai Stevenson made for a different occasion is applicable:

Let's talk sense to the American people. Let's tell them the truth, that there are no gains without pains.

A number of mechanical and chemical methods of population controls, in addition to surgical procedures, have been mentioned.

These will not work unless individuals desire them and will use them. In certain religions, such as the Catholic and the Hindu, there is opposition to these. Abortion has been common in Uruguay as was previously mentioned, and it has been said that in France there is one abortion for each live birth. A possible "carrot and stick" program that could be used to advantage within the United States would be to permit a needy mother to receive welfare maternity benefits and child support up to and through the second child. At that time, it would be explained to her very clearly that there would be no benefits after the second child, and the methods of prevention of a third birth could be clearly explained.

For persons in upper economic classes there are two methods that can be employed: one would be to permit no income tax deductions for children after the second child; another is to levy a tax on the parent for each child after the second. Still another method would be to give an allowance of a given number of years of educational scholarships to each family. If the number of years were thirty-two, this would permit the family with two children to send them through elementary school, high school, and college. If there were three children in the family, this would limit education to elementary school and high school, with the parent assuming all costs beyond that point.

At the present time, women are beginning to secure equal positions and equal pay in careers in industry, teaching, and politics. There is the possibility that this may limit the number of children spontaneously.

Many other systems of awards and punishments have been suggested. In Orwell's book *1984*, mass fertility control agents were used in food and water. This is not a system to which any group would wish to be subject. The encouragement of later marriages probably would not succeed in an age when cohabitation without marriage becomes increasingly common. That approach had worked in earlier civilizations, such as in Sparta, and perhaps since many of their warriors would now be classified as "gay," the country was not overpopulated.

There are many other ideas that might be worked out. It does not require a sheer mass of numbers of children to render Earth a pleasant place. As Wolfgang von Goethe so wisely said:

> To know of someone here and there whom we accord with, who is living on with us, even in silence—this makes our earthly ball a peopled garden.

Better inducements to population control should exist. If it could be approached from a different moral, economic, legal, and social viewpoint, much could be done. For example, if the actual facts that the population explosion is forcing upon us were widely known and understood, it would be possible to teach children that it is immoral, rather than moral, to have large families. Unless we approach a "population zero" increase rapidly, our populations will increase catastrophically. In this case, millions, or even billions, of persons will starve. Which is more moral, to regulate a population or to see huge numbers die?

There is a serious question at the present time: should the food-producing nations furnish food to those nations that take no steps to control their population? In other words, is it less moral to permit 300,000 people in a country to starve at this time, or to permit three million people in that country to starve twenty years from now?

It must be appreciated that the concern about overpopulation does not rest alone on the inability to feed vast increases in numbers of people. This is significant, of course, in light of what we have seen earlier in this book. With the laterite soils of the entire equatorial area that will not feed those nations in decades to come, the covering up of tillable lands elsewhere with the concrete trappings of civilization, and the erosion of the remaining fertile soils—together with the depletion of marine life in the oceans—famine stands as a grim specter close at hand. But it is not alone.

As hordes of people encroach ever more closely upon one another, the air becomes even more polluted with their own exhalations, coupled with the exhausts of greater numbers of vehicles and the discharges of still more factories. The effluents and wastes of increasing populations may well engulf nations in the future. Diseases spread by contact, or by water used over and over as supplies of potable water diminish, will take their toll. Each hazard, and each risk, studied previously is multiplied in geometric proportion as the population size increases arithmetically.

To make room for these additional billions will lead to the denuding of forests—badly needed as purifiers of our polluted atmosphere—as well as the loss of public preserves and park lands. Our diminishing natural resources will be depleted more rapidly. And as fossil fuels are burned more quickly, either for temperature control or for power of one type or another, the "greenhouse effect" may well be hastened, diminishing the undrowned areas available for these additional billions to dwell. Then, who is more cruel—he who says, "Be fruitful and multiply," or he who says, "Look ye to tomorrow, for ye may be without refuge"?

Is There Hope For Survival?

This is not a "scare book." It is, essentially, a dispassionate look at factors which affect the quality of life and, eventually, human survival.

Of these factors, the most immediate and present danger is that of overpopulation. If expansion should continue at the mathematical rate evidenced by the last five decades (despite the whittling effect of past wars), then the period before the epitaph is written must be shortened. Every doubling of population increases more than double the demands upon the soil, natural resources, and food and water, and intensifies the problems of waste disposal as well as disease transmission.

Even if zero population growth should be attained, our problems are not automatically solved. Our air and our waters will become more fouled, without active and dramatic changes. Food sources must be planned—of the right foods—for a century or more in the future; and progress must be planned as if each misstep could bring the day of reckoning closer.

But assuming mankind—which means the combined efforts of political, social, and religious leaders, as well as a cooperative and understanding populace—will work together toward population con-

186

trol and other measures on an international scale, then the picture becomes less grim. There would be, under those circumstances, hope for a future limited in years only by the life of the sun.

As we proceeded through this book, frequently I paused to point out methods by which certain acute problems have been solved in specific places, or to suggest possible remedies. It is not necessary to expand those now, to any degree, but some should be remembered in our further discussion. Let us, therefore, just chat—by way of summary —about some of the matters previously discussed and their importance in mankind's planning for the future.

*   *   *   *

The words of Leslie are prophetic: "We must plan our civilization or we must lose it." And I thought again of the words of Einstein, in an address delivered to a class at Cal Tech:

> It is not enough that you should understand about applied science in order that your work may increase man's blessings. Concern for man himself and his fate must always form the chief interest of all technical endeavors, concern for the great unsolved problems of the organization of labor and the distribution of goods—in order that the creations of our mind shall be a blessing and not a curse to mankind. Never forget this in the midst of your diagrams and equations.

Since man must breathe air, he must stop the ghastly pollution that is causing emphysema, lung cancer, and general debility. He could start out by not polluting himself. We hear much of the carcinogens in tobacco smoke, but how many have even stopped to think that the most minute fraction of plutonium, or any of the other long-lived members of the radioactive series, picked up by this plant from the soil or air, remains in the lung forever, and that cancer can start from this?

Let's talk about fuels for industry and for the home. With two of my colleagues, a few years ago, we proved that we could dissolve the dolomite (limestone) out of shale by using bacteria of a certain sort. The resulting fuel could be burned in situ, forming a large amount of silica slag and producing power that would run a plant. This patent was given to the University of Southern California. There is no reason why methods cannot be devised separating silica from kerogen, which can be distilled into volatile fuels like gasoline. It will require patient thought and experimentation; but the bond between kerogen and silica should be able to be broken chemically, without the need—so often discussed these days—of pulverizing the Rocky Mountains. We have reserves of kerogen in the United States that could last for 500 years.

To eliminate air pollution, if we had good control of the stacks of factories so that all fumes would be washed out, there would be little smog. The wash water could be reused, and, depending upon the type of factory, probably a useful product could be developed from the residues.

The water we drink has been adequately discussed.* If no untreated effluents were permitted to enter streams or rivers, we would have a good water supply. This would require that every factory, slaughter-house, canning plant, and other industry have its own sewage-disposal plant where adequate treatment would be given to the water before it could enter the city sewage system. If the industrial plants were small and in close proximity, they could combine in treating their effluents. If they could not meet proper standards, it would be better for them to be out of business. No chemical, solvent, pesticide, or herbicide should be permitted to enter the waters under any circumstances. Effluents should be monitored by the city, county, or state, at regular intervals, depending upon the effectiveness of treatment. This monitoring could be paid for by the company concerned; and all effluents, before entering the city sewers, should have the equivalent of secondary or acti-vated sludge treatment and a positive BOD.

With the industrial wastes pretreated as above, the sewage-treat-ment plant could give full treatment to the sewage which would consist, primarily, of untreated domestic wastes. In the near future, our secondary activated sludge liquids will be carbon filtered to absorb traces of toxic materials that otherwise might enter the towns below the plant. Sludge from anaerobic digestors can be dumped by long outfalls into deep ocean canyons along the coast. Otherwise, the sludge can be centri-fuged and dried for use as fertilizer, if there is a market, or else used for landfill.

The above has answered the questions of both water and sewage treatment except that for drinking waters that contain any turbidity or undesirable substances, they can be treated by the lime-alum or similar processes and filtered before use.

*A good article on the increasingly acute water shortage is "The Water Crisis: It's Almost Here," *Forbes*, August 26, 1979, p. 56.

The foods we eat should be as simple and as nourishing as possible. Prepared cereals containing more than 5 percent sugar should be ruled out by the FDA as, no matter what vitamins are added, sugar can produce carious lesions of the teeth. The consumer should be aware of the fact emphasized previously that, in the list of ingredients, these are indicated in descending order. In addition, the Food and Drug Administration also requires that each product have the RDA (Recommended Daily Allowance) of vitamins and minerals listed. If these are not shown, assume that they are zero.

Keep the diet of the family simple. Use complex carbohydrates such as those in yams, squash, and potatoes rather than sugared foods. Keep the diet low in fat. Be certain that the latter is done from birth on. A chubby baby may be cute to some, but he or she is manufacturing fat cells and will be a fat man or woman, or will build up fat-clogged blood vessels. Avoid between-meal snacks except on the advice of a physician. Substitute fruits for candies and your life span will probably be increased ten years. As Pope has said:

> Pleased to the last, he crops the flowery food,
> And licks the hand just raised to shed his blood.

Foods that can kill or injure us have been explained in some detail. Follow the laws of common sense, keep hot foods hot (over 145 degrees F) and cold foods cold (under 50 degrees F). Do not eat any fruits or vegetables that you cannot wash thoroughly, peel, or heat. Remember that a great many agricultural workers are legal or illegal immigrants who have high carrier rates of salmonellae, shigellae, amebic dysentery, and virus diseases.

The Food and Drug Administration is doing a good job of keeping your foods free of carcinogenic compounds. However, don't permit anyone else to use your lipstick, eyelash makeup, and certain other cosmetics that approach the eyes or mouth. Diseases we have not discussed, including infectious mononucleosis (the kissing disease), can be transmitted in this way. Do not be taken in by groups that are fighting to prevent any one certain ingredient in food. Ask yourself first, how much of this does my family use?

With reference to the "cides" that can injure or kill us, you, as an individual, are most exposed to those you use as ant or roach sprays in the home. Hold the container at arm's length and get out of the room for a while. If it kills insects, it isn't going to do you any good. The same holds true for garden sprays and, to some extent, herbicides.

Nothing needs to be added to the soils we till, for the sake of our personal health. But let me give one word of warning. Should you live in the country or in a small agricultural town, have the water tested for nitrites and nitrates particularly if there are small children in the family. You can always purchase bottled water in five-gallon containers to make up the infant's formula and for the other children. Most adults will not be injured.

We must learn to reuse our waste materials, whether it is to produce methane gas to be burned at new-type power plants or to recover the metals that we need.

The energy problem has been discussed, and, if we are sufficiently ingenious, the OPEC nations will not long have a market for their products. It will take us some years to fill in all of the gaps, but we must start now, not tomorrow. This problem includes the "nukes" we build, but, although I gave all of the bad points, I am in favor of the use of nuclear power. Unfortunately, we started too soon before we had the technology developed for caging the savage beast.

In Gratia and In Memoriam

With utmost humility, to the three most important persons of my life: my wife, Lucille E. Appleman, who has struggled through the first copies of my many manuscripts and books; our son, Milo Don Appleman, Jr., M.D., now deceased, with whom I formerly discussed important problems in medicine; and my brother, John Alan Appleman, the author of many books, who has been my guide and mentor in the writing of this volume.

My thanks go also to Dr. Harrison M. Kurtz, Dr. John Luther Mohr, Dr. Louis C. Wheeler, and Theodore Bertness, whose great funds of knowledge were helpful in every aspect of this book; and to Bethel Brinton for helping to bring these materials together into a readable unit.